Progress in Mathematics
Vol. 24

Edited by
J. Coates and
S. Helgason

Birkhäuser
Boston · Basel · Stuttgart

Enumerative Geometry and Classical Algebraic Geometry

Patrick Le Barz,
Yves Hervier,
editors

1982

Birkhäuser
Boston • Basel • Stuttgart

Patrick Le Barz
Université de Nice
Département de Mathématiques
Parc Valrose
F-06034 Nice, France

Yves Hervier
Université de Nice
Déparement de Mathématiques
Parc Valrose
F-06034 Nice, France

Library of Congress Cataloging in Publication Data
Main entry under title:

Enumerative geometry and classical algebraic geometry.
(Progress in mathematics ; v. 24)
1. Geometry, Algebraic--Addresses, essays, lectures.
2. Fields, Algebraic-Addresses, essays, lectures.
3. Algebraic varieties--Addresses, essays, lectures.
I. Le Barz, Patick, 1948- . II. Hervier, Y.
III. Series: Progress in mathematics (Cambridge, Mass.) ;
v. 24.
QA564.E58 1982 516.3'5 82-17748
ISBN-13: 978-0-8176-3106-2 e-ISBN-13: 978-1-4684-6726-0
DOI: 10.1007/978-1-4684-6726-0

CIP - Kurztitelaufnahme der Deutschen Bibliothek

LeBarz, Patrick:

Enumerative geometry and classical algebraic
geometry / Patrick Le Barz and Yves Hervier. -
Boston; Basel; Stuttgart : Birkhäuser, 1982.
(Progress in mathematics; Vol. 24)
ISBN-13: 978-0-8176-3106-2

NE: Hervier, Yves:; GT

Softcover reprint of the hardcover 1st edition 1982

ISBN-13: 978-0-8176-3106-2

TABLE DES MATIERES

AUTRES COMMUNICATIONS FAITES AU COURS DU COLLOQUE

E. BALLICO	Postulation formulae for generic elliptic curves.
M. FIORENTINI	Trois théorèmes sur les courbes de $\mathbb{P}^3$.
F. GAETA	Formes de Cayley-Chow de l'intersection de deux variétés algébriques.
K. HULEK	The normal bundle of a space curve on a quadric.
H. LANGE	Vector bundles on curves and secant planes.
R. LAZARSFELD	Varieties with degenerate secant varieties.
G. MARTENS	On the Clifford-index of a curve.
U. PERSSON	Surfaces with maximal Picard numbers.
G. SACCHIERO	Les variétés paramétrisant les courbes rationnelles de $\mathbb{P}^3$, de fibré normal fixé.
E. STAGNARO	On Basset's formulae for the maximum number of nodes of a surface.
G. WÜSTHOLZ	Divisors on algebraic groups.

PREFACE

Ce volume est formé de la version définitive des textes des conférences faites à Nice au cours d'un Colloque qui s'y est tenu du 23 au 27 Juin 1981.

Comme le suggère son titre, le sujet, volontairement restreint, gravitait grosso-modo autour des espaces projectifs de petite dimension et des courbes, cela sur un corps algébriquement clos.
Il semble que ce choix délibéré ait été bien accueilli dans l'ensemble par les participants.

Le Colloque a rassemblé une soixantaine de spécialistes venus de différents pays et nous nous sommes en particulier réjouis de l'importante participation de nos voisins italiens.

Le professeur DIEUDONNE n'ayant pas pu participer au colloque comme il était initialement prévu, a cependant bien voulu accepter la charge ingrate d'organiser le "refeering" de tous les textes présentés ici. Nous l'en remercions très vivement.

En plus de leur compétence, l'organisation matérielle a bénéficié des sourires de Mesdames Heike Laurin et Annie Borel.

La maison Birkhaüser a bien voulu accepter de publier les Actes du Colloque dans sa collection Progress in Mathematics, avec la diligence et le soin que l'on connaît.

Enfin, l'existence même du colloque a été rendue possible, grâce aux subventions des organismes ci-dessous :

- Société Mathématique de France.
- Conseil Général des Alpes-Maritimes.
- Comité Lépine de la ville de Nice.
- Université de Nice.
- Ecole des Mines de Paris.
- Département de Mathématiques de l'I.M.S.P.

Qu'ils en soient tous très chaleureusement remerciés.

Les organisateurs, Y. HERVIER & P. LE BARZ.

LISTE DES PARTICIPANTS

NOM, Prénom	Université	Pays
BALLICO Edoardo	Scuola Normale Sup., Pisa	Italie
BAPTISTA de CAMPOS Manuel	Nice	France
BEAUVILLE Arnaud	Ecole Polytechnique Palaiseau	France
BERNASCONI Carlo	Perugia	Italie
BERTRAND Daniel	Paris VI	France
BERTIN José	Toulouse III	France
BOUTOT Jean-François	Strasbourg	France
BRUN Jerôme	Nice	France
CATANESE Fabrizio	Pisa	Italie
CHIARLI Nadia	Torino	Italie
CLEMENS Charles Herbert	Salt Lake City, UT	U.S.A.
COLLINO Alberto	Torino	Italie
COPPENS M.	Utrecht	Pays Bas
COX David	Amherst, MA	U.S.A.
CUMINO Caterina	Torino	Italie
DALALLIAN Samuel	Paris	France
DANA PICARD Thierry	Nice	France
DAOUDI Mohamed	Nice	France
DE BOER Jan H.	Nijmegen	Pays Bas
DEL CENTINA Andrea	Firenze	Italie
DESCHAMPS Mireille	Orsay	France
DLOUSSKY Georges	Marseille	France
ELENCWAJG Georges	Nice	France
ELLIA Philippe	Nice	France
ELLINGSRUD Geir	Nice	France
EYSSETTE Frédéric	Nice	France
FERRAND Daniel	Paris	France
FERRARESE Giorgio	Torino	Italie

FINAT CODES Francisco	Valladolid	Espagne
FIORENTINI Mario	Ferrare	Italie
FORSTER Otto	Münster	Allemagne
FULTON William	Princeton, N.J.	U.S.A.
GAETA Federico	Barcelone	Espagne
GALLIGO André	Nice	France
GIANNI Patrizia	Pisa	Italie
GRUSON Laurent	Lille	France
HARBATER David	PHiladelphia, Penn.	U.S.A.
HEFES Abramo	Firenze	Italie
HERNANDEZ Daniel	Salamanca	Espagne
HERNANDEZ Garcia Rafael	Tafala	Navarro
HERVIER Yves	Nice	France
HIRONAKA Heisuke	Cambridge, MA	U.S.A.
HIRSCHOWITZ André	Nice	France
HULEK Klaus	Erlangen	Allemagne
KLEIMAN Steve	MIT, Cambridge, MA	U.S.A.
LAKSOV Dan	Stockholm	Suède
LANGE Herbert	Erlangen	Allemagne
LAZARSFELD Robert	Cambridge, MA	U.S.A.
LE BARZ Patrick	Nice	France
LEVINE Marc	Philadelphia, Penn.	U.S.A.
MAISONOBE Philippe	Nice	France
MARTENS Gerriet	Erlangen	Allemagne
MENEGAUX Renée	Orsay	France
MITSCHI Claude	Strasbourg	France
MORVAN Jean-Marie	Toulouse	France
NARASIMHAN M.S.	Tata Inst., Bombay	Inde
NAVARRO, Juan A.	Salamanca	Espagne
PARADIS Denis	Montréal	Canada
PARADIS Louise	Montréal	Canada
PARIGI Giuliano	Ferrara	Italie
PAXIA Giuseppe	Catania	Italie
PESKINE Christian	Oslo	Norvège
PHAM Frédéric	Nice	France
PHILIPPON P.	Palaiseau	France
PIENE Ragni	Oslo	Norvège
REGOLI G.	Perugia	Italie

SACCHIERO Gianni	Oslo	Norvège
SCHNEIDER Michael	Bayreuth	Allemagne
SPANGHER Walter	Trieste	Italie
STAGNARO Ezio	Padova	Italie
TRAVERSO Carlo	Pisa	Italie
VAINSENCHER Israel	Recife, PE	Brésil
VERSCHOREN Alain	Wilrijk	Belgique
VARGAS J.A.	Mexico	Mexique
WALLACE David	Cambridge, MA	U.S.A.
WARDELMANN R.	Göttingen	Allemagne
WÜSTHOLZ Gisbert	Wuppertal	Allemagne

COURBES DE L'ESPACE PROJECTIF: VARIETES DE SECANTES

par

Laurent GRUSON et Christian PESKINE

La rédaction présentée ici a pour origine une tentative de réflexion sur les variétés de droites sauteuses (ou multi-sauteuses) d'un fibré stable de rang deux sur P^3 (1). Comme celles-ci, les variétés de sécantes d'une courbe localement intersection complète (de P^3) sont localement définies, dans la grassmannienne des droites, par les mineurs maximaux de matrices persymétriques (i.e. de la forme (u_{i+j})). Bien sûr, dans les deux cas, l'énumération se fait par la formule de Porteous-Kempf-Laksov, lorsque les codimensions sont les bonnes. Dans l'étude locale, les difficultés non surmontées pour les droites sauteuses sont aisément résolues pour les variétés de sécantes à une courbe lisse. Nous développons donc une description de propriétés locales et globales de ces dernières, naturellement plongées dans la grassmannienne au dessus de laquelle on raisonne par élimination à une variable.

Le point de vue décrit dans le premier paragraphe diffère de celui des auteurs modernes qui préfèrent travailler dans les variétés de diviseurs sur la courbe (ce qui comporte des avantages évidents pour qui juge artificiel de considérer une quadrisécante comme une trisécante singulière).

L'étude globale est faite dans le deuxième paragraphe où nous vérifions les formules énumératives classiques: degré et genre arithmétique de la courbe des trisécantes, nombre de quadrisécantes. La méthode repose sur l'utilisation de propriétés des complexes d'Eagon-Northcott que nous rappelons. Ceci apparait particulièrement ensuite dans le

(1) Espace projectif de dimension 3 sur un corps algébriquement clos de caractéristique nulle.

calcul du genre de la courbe obtenue par éclatement de la variété des quadrisécantes (lorsqu'elle est finie) dans la courbe des trisécantes. Dans le troisième paragraphe (où la courbe de P^3 considérée est toujours supposée lisse) nous calculons la dimension de l'espace tangent à la variété des n-sécantes en un point de la grassmannienne et donnons une condition nécessaire et suffisante, explicite, pour que l'éclatement de la variété des quadrisecantes (lorsqu' elle est finie) dans la courbe des trisécantes soit non-singulier Le lecteur indulgent remarquera l'adjonction de deux appendices. Dans le premier, on voit que pour l'étude des adjointes d'ordre supérieur (à une courbe tracée sur une surface lisse) le point de vue des idéaux de Fitting épouse parfaitement celui des points infiniment voisins. Dans le second, la méthode décrite dans notre deuxième paragraphe est reprise pour étudier les variétés de droites multisauteuses d'un fibré stable de rang 2 sur P^n; quelques calculs sont faits explicitement (le principe a déjà été utilisé par Barth ([1]), ainsi que dans un manuscrit non publié de J. Bertin).

§ 1. Sécantes à une courbe de P^3

Soit C une courbe de P^3 ne contenant pas de droite. Notons G la grassmannienne des droites de P^3 et F la variété d'incidence dans $G \times P^3$. Si q et p sont les applications naturelles de F dans G et P^3, pour tout θ_F-module M nous écrivons $M(n,m)$ pour

$M \otimes_{\theta_F} q^*\theta_G(n) \otimes_{\theta_F} p^*\theta_{P^3}(m)$, où $\theta_G(1)$ est le générateur ample de Pic G. Soit J le faisceau d'idéaux de l'image réciproque $\mathcal{C}$ de C dans F. Pour $k \gg 0$, on a $R^1q_*J(0,k) = 0$; la suite

$0 \to q_*J(0,k) \to q_*\theta_F(0,k) \to q_*\theta_{\mathcal{C}}(0,k) \to 0$ est alors exacte et $q_*J(0,k)$ est un fibré de rang $k+1$ sur G.

Définition 1.1: <u>Soit</u> i <u>un entier positif; l'idéal de la variété des i-sécantes à</u> C <u>est le</u> $(i-1)$<u>-ième idéal de Fitting du</u> θ_G<u>-module</u> $q_*\theta_{\mathcal{C}}$.

Comme $q_*\theta_{\mathcal{C}}(0,k) = q_*\theta_{\mathcal{C}}(k,0)$, il est clair que cet idéal est le $(i-1)$-ième idéal de Fitting de la présentation $q_*J(0,k) \to q_*0_F(0,k)$ pour $k \gg 0$.

Proposition 1.2: <u>Pour</u> $i \geq 1$, <u>et pour</u> $0 \leq r \leq i-1$, <u>la variété des</u> i-<u>sécantes à</u> C <u>est définie par le</u> r-<u>ième idéal de Fitting de</u> $R^1q_*J(0,i-2-r)$.

Pour $r = i-1$ c'est la définition, compte tenu de l'isomorphisme $R^1q_*J(0,-1) \simeq q_*\theta_{\mathcal{C}}(-1,0)$. Pour $r \leq i-2$, considérons R l'anneau d'un ouvert affine U de G et A l'anneau de l'ouvert affine de $\mathcal{C}$ image réciproque de U. Comme A est fini monogène sur R, la proposition est un cas particulier du résultat suivant:

Lemme 1.3: <u>Soit</u> $A = R[x]$ <u>un anneau entier sur</u> R <u>engendré par</u> $1,x,\ldots,x^k$ <u>(comme</u> R-<u>module). Si</u> A_ℓ <u>est le sous</u> R-<u>module de</u> A <u>engendré par</u> $1,x,\ldots,x^\ell$, <u>le</u> 0-<u>ième idéal de Fitting de</u> A/A_{k-s} <u>est égal au</u> $(k+1-s)$-<u>ième idéal de Fitting de</u> A.

Soit $P(X)$ un polynôme unitaire de degré $k+1$ appartenant à l'idéal de A dans $R[X]$. Notons $\bar{X}$ l'image de X dans $V = R[X]/(P(X))$. Rappelons que $\mathrm{Hom}_R(V,R)$ est un V-module libre de rang 1 engendré par τ où $\tau(\bar{X}^i) = \delta_{i,k}$ pour $0 \leq i \leq k$. Si $a_0,\ldots,a_{s-1}$ sont des éléments de l'idéal I de A dans V, l'idéal qu'ils définissent dans V est engendré comme R-module par $X^m a_i$ où $i \in [0,s-1]$ et $m \geq 0$. Donc le $(k+1-s)$-ième idéal de Fitting de A est engendré par les éléments $\det\,[X^{n_j}\tau(X^{m_i}a_i)]$ où $a_0,\ldots,a_{s-1} \in I$, où $m_i \geq 0$ pour $0 \leq i \leq s-1$, et où $0 \leq n_0 < n_1 < \cdots < n_{s-1} \leq k$. On vérifie alors que l'orthogonal du sous R-module libre de V engendré par $1,\bar{X},\ldots,\bar{X}^{k-s}$ est engendré par $X^j\tau$ avec $0 \leq j \leq s-1$. On en déduit que le 0-ième idéal de Fitting de A/A_{k-s} est engendré par les éléments $\det\,[X^j\tau(X^{m_i}a_i)]$ avec $0 \leq j \leq s-1$. Posant $\tau(\bar{X}^m a_i) = \alpha_{i,m}$, le lemme est conséquence du résultat suivant:

Lemme 1.4: <u>Soient</u> $(\alpha_{i,n})$ $0 \leq i \leq s-1$ <u>et</u> $n \geq 0$ <u>des suites d'éléments d'un anneau</u> R. <u>Pour</u> $m_i \geq 0$ <u>et</u> $0 \leq n_0 < n_1 < \cdots < n_{s-1}$, <u>posons</u> $D_{n_0,\ldots,n_{s-1}}^{m_0,\ldots,m_{s-1}} = \det(\alpha_{i,m_i+n_j})$. <u>Alors</u> $D_{n_0,\ldots,n_{s-1}}^{m_0,\ldots,m_{s-1}}$ <u>est combinaison linéaire à coefficients dans</u> $\mathbb{Z}$ <u>des éléments</u> $D_{0,\ldots,s-1}^{r_0,\ldots,r_{s-1}}$ <u>tels que</u> $r_i+s-1 \leq m_i+n_{s-1}$.

Considérons s indéterminées $Y_0,\ldots,Y_{s-1}$. On sait (déterminants de Vandermonde généralisés) qu'il existe une identité $\det(Y_i^{m_i+n_j}) = \det(Y_i^j)\cdot\sum\lambda_{r_0,\ldots,r_{s-1}} Y_0^{r_0}\times\ldots\times Y_{s-1}^{r_{s-1}}$, soit $\det(Y_i^{m_i+n_j}) = \sum \lambda_{r_0,\ldots,r_{s-1}} \det(Y_i^{r_i+j})$ où $\lambda. \in \mathbb{Z}$ et $\lambda_{r_0,\ldots,r_{s-1}} \neq 0$ implique $r_i+s-1 \leq m_i+n_{s-1}$. Notons maintenant $\Delta^{m_0,\ldots,m_{s-1}}_{n_0,\ldots,n_{s-1}}$ l'opérateur différentiel $\text{dét}(\partial^{m_i+n_j}_{\bullet}/\partial Y_i^{m_i+n_j})$. On a évidemment

$$\Delta^{m_0,\ldots,m_{s-1}}_{n_0,\ldots,n_{s-1}} = \sum \lambda_{r_0,\ldots,r_{s-1}} \Delta^{r_0,\ldots,r_{s-1}}_{0,\ldots,s-1}.$$

Introduisons enfin les fonctions

$f_i(T) = \sum_{n\geq 0}(\alpha_{i,n}/n!)T^n$. On vérifie

$$\Delta^{m_0,\ldots,m_{s-1}}_{n_0,\ldots,n_{s-1}}(f_0(Y_0)\times f_1(Y_1)\times\ldots\times f_{s-1}(Y_{s-1}))(0,0..,0)$$

$= D^{m_0,\ldots,m_{s-1}}_{n_0,\ldots,n_{s-1}}$, ce qui démontre le lemme.

Montrons que cette définition des variétés de sécantes (probablement bien connue mais pour laquelle nous n'avons pas trouvé de référence) est compatible avec l'intuition courante en ce qui concerne les bisécantes:

Considérons $P^2 \subset G$ le plan formé des droites passant par un point qui n'est pas le sommet d'un cône de multisécantes à C. Il est clair que la restriction de $R^1q_*J(0,0)$ au plan P^2 choisi est le conoyau de l'application $\theta_{\bar{C}} \to \theta_C$ où $\bar{C}$ est la projection de C dans P^2. L'idéal de Fitting se restreignant évidemment, et le conducteur de θ_C dans $\theta_{\bar{C}}$ étant la trace de θ_C, consideré comme $\theta_{\bar{C}}$-module, il suffit pour vérifier que l'intersection de notre variété de bisécantes avec P^2 est la variété définie par le conducteur de 0_C dans $0_{\bar{C}}$ de démontrer l'énoncé suivant.

Lemme 1.5: <u>Soit</u> M <u>un module de Cohen-Macaulay de dimension</u> d-1, <u>de type fini sur un anneau régulier local</u> R <u>de dimension</u> d, <u>et soit</u> $0 \to R^s \overset{B}{\to} R^s \to M \to 0$ <u>une résolution de M; alors la trace du</u> $[R/(\text{dét}B)]$-<u>module</u> M <u>est le premier idéal de Fitting de</u> M.

Considérons le diagramme commutatif suivant:

$$\begin{array}{ccccccccc}
0 & \longrightarrow & (R^s)^{\vee} & \xrightarrow{\times\ \text{dét } B} & (R^s)^{\vee} & \longrightarrow & (R^s)^{\vee}\otimes_R R/(\text{dét } B) & \longrightarrow & 0 \\
 & & \downarrow t_B & & \downarrow t_B & & \downarrow t_B \otimes_R R/(\text{dét } B) & & \\
0 & \longrightarrow & (R^s)^{\vee} & \xrightarrow{\times\ \text{dét } B} & (R^s)^{\vee} & \longrightarrow & (R^s)^{\vee}\otimes_R R/(\text{dét } B) & \longrightarrow & 0
\end{array}$$

Explicitons l'isomorphisme $\mathrm{Ext}^1_R(M,R) \overset{\pi}{\cong} \mathrm{Hom}_R(M,R/(\det B))$ donné par le diagramme du serpent. Pour cela, fixons deux bases (f_j) et (e_i) de R^s et soit (α_{ij}) la matrice de B dans ces bases, $B(f_j) = \sum \alpha_{ij}\, e_i$. On vérifie alors directement, en notant (f_j^v) et (e_i^v) les bases duales, que la classe de f_j^v dans $\mathrm{Ext}^1_R(M,R)$ a pour image la classe de $\sum_i A_{ij}\, e_i^v$ dans $(R^s)^v \otimes_R R/(\det B)$, où A_{nm} est le cofacteur de α_{nm} ; le lemme s'en déduit immédiatement.

§ 2. Enumération des sécantes

Faisons au prélable quelques rappels (d'aprés Kempf) sur les complexes d'Eagon-Northcott.

Considérons Y une variété lisse quasi-projective et $L \xrightarrow{\phi} N$ une application linéaire de fibrés vectoriels sur Y, de rangs respectifs ℓ et n. Notons $P = \mathrm{Proj}\ (\mathrm{sym}\ N)$ et $\theta_P(1)$ le module inversible sur P ample au dessus de Y. Il y a un complexe de Koszul

$$K. \quad 0 \to \Lambda^{\ell}(L_P(-1)) \to \cdots \to L_P(-1) \to 0_P ,$$

où L_P désigne l'image réciproque de L sur P. Pour $-1 \leqslant i \leqslant \ell-n+1$, la suite spectrale d'hypercohomologie du foncteur image directe sur Y appliquée à $K.\otimes \theta_P(i)$ dégénère en un complexe E_i

$$0 \to \Lambda^n N^v \otimes \Lambda^{\ell} L \otimes S_{\ell-n-i}(N^v) \to \cdots \to \Lambda^n N^v \otimes \Lambda^{n+i} L \to \Lambda^i L \to \cdots$$
$$\cdots \to L \otimes S_{i-1}(N) \to S_i(N) \quad ,$$

où $S_k(Q)$ est la k-ième puissance symétrique du module Q. Soit X la variété des zéros de $\Lambda^n\phi$; l'homologie des complexes E_i est à support dans X. Rappelons que $\mathrm{codimension}_Y(X) \leqslant \ell-n+1$, et supposons à partir de maintenant que ces nombres sont égaux. Comme les complexes E_i sont de longueur $\ell-n+1$, ils sont exacts. Si M_i est le θ_X-module dont E_i est une résolution, on a $M_0 = \theta_X$ et $M_i = S_i(\mathrm{coker}\ \phi)$ pour $i \geqslant 1$.

Proposition 2.1: (I) <u>Pour tout point</u> $x \in X$ <u>de codimension</u> $\ell-n+1$ <u>dans</u> Y, <u>la longueur de</u> $(M_i)_x$ <u>est indépendante de</u> i, <u>avec</u> $-1 \leq i \leq \ell-n+1$.

(II) <u>Lorsque</u> Y <u>est projective, la variété</u> X <u>représente dans l'anneau de Chow de</u> Y <u>la classe de Chern</u> $c_{\ell-n+1}$ $(\operatorname{coker} \phi^{\vee})$.

(III) <u>Le</u> θ_X-<u>module</u> $\omega_X = M_{\ell-n} \otimes_{\theta_Y} \omega_Y \otimes_{\theta_Y} \Lambda^n N \otimes_{\theta_Y} \Lambda^{\ell} L^{\vee}$ (<u>où</u> ω_Y <u>est un</u> θ_Y-<u>module dualisant) est dualisant</u>. <u>De plus</u>, $\operatorname{Hom}_{\theta_X}(M_i, M_{\ell-n}) \simeq M_{\ell-n-i}$ <u>pour</u> $-1 \leq i \leq \ell-n+1$.

(IV) <u>Il existe une application naturelle</u> $M_{-1} \otimes_{\theta_X} M_1 \to \theta_X$ <u>ayant pour image la restriction à</u> θ_X <u>du premier idéal de Fitting de</u> ϕ; <u>elle identifie</u> M_{-1} <u>à un facteur direct de</u> $\operatorname{Hom}_{\theta_X}(M_1, \theta_X)$.

Posons $R = \theta_{Y,x}[T_{ij}]_{\substack{1 \leq i \leq n \\ 1 \leq j \leq \ell}}$. Notons Φ l'application linéaire $R^{\ell} \to R^n$ de matrice (T_{ij}). Effectuons la construction de Kempf pour Φ et notons $\mathcal{M}_i$ $(-1 \leq i \leq \ell-n+1)$ les modules résolus par les complexes d'Eagon-Northcott de Φ. Soit (a_{ij}) une matrice de $(\phi)_x$; identifions $\theta_{Y,x}$ au quotient $R/(T_{ij}-a_{ij})$. Si $A = \mathcal{M}_0$ est l'anneau de la variété des zéros de $\Lambda^n \Phi$, il est clair que A est intègre et que $\mathcal{M}_i$ est un A-module sans torsion de rang 1, de Cohen-Macaulay. Les longueurs considérées dans (I) sont donc toutes égales à la multiplicité d'intersection des R-modules A et $\theta_{Y,x}$ au point $x \in \operatorname{Spec} \theta_{Y,x} \subset \operatorname{Spec} R$.

(II) est le cas le plus simple de la formule de Porteous démontrée par Kempf et Laksov (voir par exemple [5]).

(III) se démontre en remarquant que le transposé du complexe $E_i \otimes_{\theta_Y} \Lambda^n N \otimes_{\theta_Y} \Lambda^{\ell} L^{\vee}$ est $E_{\ell-n-i}$ et en rappelant que le produit tensoriel avec ω_Y du transposé de E_0 est une résolution d'un θ_X-module dualisant.

Pour démontrer (IV), considérons les présentations données de M_1 et M_{-1}:

$L \to N \to M \to 0$, et $\Lambda^n N^{\vee} \otimes_{\theta_Y} \Lambda^n L \otimes_{\theta_Y} N^{\vee} \to \Lambda^n N^{\vee} \otimes_{\theta_Y} \Lambda^{n-1} L \to M_{-1} \to 0$.

L'isomorphisme $(\Lambda^{n-1}L \otimes_{\theta_Y} \Lambda^n N^v) \otimes_{\theta_Y} N \to \Lambda^{n-1}L \otimes_{\theta_Y} \Lambda^{n-1}N^v$ induit une application naturelle $(\Lambda^n N^v \otimes_{\theta_Y} \Lambda^{n-1}L) \otimes_{\theta_Y} N \to \theta_Y$ dont l'image est le 1er idéal de Fitting de ϕ. On vérifie facilement que les deux applications composées

$$(\Lambda^n N^v \otimes_{\theta_Y} \Lambda^{n-1}L) \otimes_{\theta_Y} L \to (\Lambda^n N^v \otimes_{\theta_Y} \Lambda^{n-1}L) \otimes_{\theta_Y} N \to \theta_Y$$

$$(\Lambda^n N^v \otimes_{\theta_Y} \Lambda^n L \otimes_{\theta_Y} N^v) \otimes_{Q_Y} N \to (\Lambda^n N^v \otimes_{\theta_Y} \Lambda^{n-1}L) \otimes_{\theta_Y} N \to \theta_Y$$

sont à valeurs dans le 0-ième idéal de Fitting de ϕ. Définissons maintenant une application linéaire $\mathrm{Hom}_{\theta_X}(M_1, \theta_X) \to M_{-1}$. Nous avons vu (III) qu'il existe un isomorphisme naturel $M_{-1} \simeq \mathrm{Hom}_{\theta_X}(M_{\ell-n+1}, M_{\ell-n})$; mais une forme linéaire sur M_1 se prolonge de façon unique en une dérivation homogène de degré -1 de $\mathrm{sym}.(M_1)$ et induit donc une application linéaire $M_{\ell-n+1} \to M_{\ell-n}$. Montrons que l'application composée $M_{-1} \to \mathrm{Hom}_{\theta_Y}(M_1, \theta_X) \to M_{-1}$ est un isomorphisme. Il suffit évidemment de le faire dans le cas "générique". Mais dans ce cas les applications considérées sont des isomorphismes hors de la variété des zéros de $\Lambda^{n-1}\phi$; comme celle-ci est de codimension ≥ 2 dans la variété des zéros de $\Lambda^n\phi$, on conclut en remarquant de plus que les modules concernés sont réflexifs.

Proposition 2.2: <u>Si</u> Y <u>est affine, et s'il existe des bases de</u> L <u>et</u> N <u>dans lesquelles</u> ϕ <u>admet une matrice persymétrique il y a un homomorphisme surjectif</u> $\mathrm{sym}_{\ell-n+2}(M_1) \to \mathfrak{a}_1$ <u>où</u> $\mathfrak{a}_1$ <u>est le premier idéal de Fitting du</u> θ_X<u>-module</u> M_1.

On peut supposer $Y = \mathrm{Spec}\; k[(U_i)0 \leq i \leq \ell+n-2]$ et $\phi = (U_{i+j})_{\substack{0 \leq i \leq n-1 \\ 0 \leq j \leq \ell-1}}$ où les U_i sont des indéterminées.

Posons $A = \theta_X$, et considérons l'anneau gradué $B = \oplus \bar{M}_i$ où $\bar{M}_i$ est le A-module sans torsion de rang un quotient de $\mathrm{sym}_i(M)$. Soit $(e_0, \ldots, e_{n-1})$ le système de générateurs de M_1 image de la base canonique de N. Notant $\bar{U}_i$ l'image de U_i dans A, il est clair que le produit suivant de matrices à coefficients dans B est nul:

$$\begin{pmatrix} \bar{U}_0 & ,\dots, & \bar{U}_\ell \\ \vdots & & \vdots \\ \bar{U}_{n-2} & ,\dots, & \bar{U}_{\ell+n-2} \end{pmatrix} \times \underbrace{\begin{pmatrix} e_0 & 0 & \dots & 0 \\ \vdots & & & \vdots \\ & & & e_0 \\ e_{n-1} & & & \vdots \\ 0 & & & \vdots \\ \vdots & & & \\ 0 & & & e_{n-1} \end{pmatrix}}_{\ell-n+2 \text{ colonnes}}$$

D'après (Bourbaki, Algèbre, chap.3, §8, ex.13) il existe un isomorphisme homogène de degré $\ell-n+2$ entre le 0-ième idéal de Fitting de la première matrice et celui de la seconde; il induit clairement un isomorphisme de A-modules entre le 0-ième idéal de Fitting de la matrice $(\bar{U}_{i+j})_{\substack{0\le i\le n-2\\ 0\le j\le \ell}}$ à coefficients dans A, et $\bar{M}_{\ell-n+2}$. La proposition est alors conséquence du résultat suivant:

Lemme 2.3: Soit u_i $(i\geq 0)$ une suite infinie d'éléments d'un anneau. Pour $n \le \min(p,q)$, l'idéal engendré par les n-mineurs de la matrice $N_{p,q} = (u_{i+j})_{\substack{0\le i\le p-1\\ 0\le j\le q-1}}$ ne dépend que de $p+q$.

Soient $(X_1,\dots,X_n)$ des indéterminées. Pour toute partie I à n éléments de N, on pose $V_I = \text{dét}(X_i^{m_j})$ où $m_1 < m_2 < \dots < m_n$ est la suite croissante des éléments de I.

Si $(r_1,\dots,r_n)$ est une suite d'entiers, on a

$\text{dét}(X_i^{r_i+m_j}) = X_1^{r_1}\times\dots\times X_n^{r_n}\times V_I$, donc

$\sum_{\sigma\in S_n} \det(X_{\sigma(i)}^{r_1+m_j}) = \sum_{\sigma\in S_n} X_{\sigma(1)}^{r_1}\times\dots\times X_{\sigma(n)}^{r_n} V_I(X_{\sigma(1)},\dots,X_{\sigma(n)})$.

Comme V_I est une fonction antisymétrique des X_i, cette dernière expression est aussi $\sum_{\sigma\in S_n} \varepsilon(\sigma) X_{\sigma(1)}^{r_1}\times\dots\times X_{\sigma(n)}^{r_n} V_I$.

Lorsque r_i est la suite strictement croissante associée à une partie J à n éléments de N, on trouve

$\sum_{\sigma\in S_n} \text{dét}(X_{\sigma(i)}^{r_i+m_j}) = V_I \cdot V_J$.

Maintenant, si $T_1,\ldots,T_n$ sont des indéterminées, considérons les opérateurs différentiels $D_I = \text{dét}(\partial^{m_j} \cdot / \partial T_i^{m_j})$ et $D_{I,J} = \text{dét}(\partial^{r_i+m_j}/\partial T_i^{r_i+m_j})$. Si s est une fonction symétrique de $(T_1,\ldots,T_n)$, on a démontré

$n!D_{I,J}(s) = D_I \circ D_J(s)$. Considérons alors la fonction

$s(T_1,\ldots,T_n) = f(T_1)\times f(T_2)\times\ldots\times f(T_n)$, où $f(X) = \sum_{i\geq 0} u_i X^i/i!$,

et remarquons que $D_{I,J}(s)(0,\ldots,0) = \begin{vmatrix} u_{r_1+m_1}, \ldots, u_{r_1+m_n} \\ \vdots \qquad \vdots \\ u_{r_n+m_1}, \ldots, u_{r_n+m_n} \end{vmatrix}$.

Supposons $r_n+m_n \leq p+q-2$, nous voulons montrer que ce déterminant est combinaison linéaire de n-mineurs de la matrice $N_{p,q}$. Pour cela, rappelons que les fonctions de Schur $\tau_I = V_I/V_{[0,n-1]}$ forment pour $I \subset [0,k]$ une base des fonctions symétriques de $(X_1,\ldots,X_n)$ de degré $\leq k-n+1$ (en les fonction symétriques fondamentales); on en déduit que $\tau_I \cdot \tau_J = \sum_{\substack{K\subset[0,p-1] \\ L\subset[0,q-1]}} \alpha_{K,L} \tau_K \cdot \tau_L$. Bien évidemment ceci démontre le lemme puisqu'on a alors

$D_{I,J}(s)(0) = \sum_{\substack{K\subset[0,p-1] \\ L\subset[0,q-1]}} \alpha_{K,L} D_{K,L}(s)(0)$.

Venons en à l'étude des classes, dans l'anneau de Chow de la grassmannienne, des variétés de sécantes d'une courbe C de P^3 ne contenant pas de droite. On sait que l'anneau de Chow de G est $Z[t,u]$ avec $d^o t = 1$ et $d^o u = 2$, les relations étant $t^3 = 2t\cdot u$ et $u^2 = t^2 u$; ici t est representé par les droites d'un complexe linéaire et u par les droites d'un plan. Pour $r \leq 4$, supposons que la variété des r-sécantes a codimension r dans G. Montrons dans ce cas, comment sa classe se calcule naturellement au moyen du théorème de Riemann-Roch-Grothendieck, compte tenu de 2.1 (II).

Considérons une présentation $0 \to \Sigma' \to \Sigma \to J(r-2) \to 0$ où J est l'ideal de C dans P^3 et où Σ' et Σ sont des fibrés sur P^3 tels que $h^o(\Sigma \otimes_{\theta_{P^3}} 0_L) = 0$ pour toute droite L.

Notons $0 \to E' \to E \to J(0,r-2) \to 0$ l'image réciproque de cette présentation sur F. Elle induit une suite exacte

$$0 \to q_* J(0,r-2) \to R^1 q_* E' \overset{\phi}{\to} R^1 q_* E \to R^1 q_* J(0,r-2) \to 0.$$

La codimension du support de $R^1 q_* J(0,r-2)$ est r; les θ_G-modules $R^1 q_* E$ et $R^1 q_* E'$ sont localement libres et la différence de leurs rangs est le rang de $q_* J(0,r-2)$, soit r-1. D'aprés (2.1 (II)), la classe de la variété des r-sécantes est $c_r(\text{coker}\ \phi^v) = (-1)^r c_r(q!J(0,r-2))$. La méthode étant claire, donnons les étapes du calcul pour r = 3,4. Pour cela, notons $Z[X]/(X^4) = Z[x]$ l'anneau de Chow de P^3. On a alors

$c_t(J(k)) = 1+kx+dx^2+[2g-2+d(4-k)]x^3,$

où c_t est la classe de Chern totale; bien entendu, d et g désignent le degré et le genre arithmétique de C.

$ch(J(1)) = 1+x+[(1-2d)/2]x^2+[(1+6g-6+6d)/6]x^3$

$ch(J(2)) = 1+2x+(2-d)x^2+[(3g+1)/3]x^3$

où ch est le caractère de Chern. L'anneau de Chow de F s'identifie au produit tensoriel sur Z des anneaux de Chow de G et P^3 modulo la relation $x^2-tx+u = 0$; la classe de Todd du morphisme $F \to G$ est

$\tau = 1+(2x-t)/2+(t^2-4q)/12-q^2/72$. D'aprés R-R-G, on a alors

$ch(q!J(0,1)) = q_*(chJ(0,1)\times\tau)$

$= 2+(1-d)t+([2g+d-1)/2]t^2-gu)-[(3g+2d-2)/6]tu+[(g+d-1)/12]u^2$

et

$ch(q!J(0,2)) = 3+(3-d)t+([2g+3-d)/2]t^2-[g+3-d]u)$

$-[(3g+3-d)/6]tu+(g/12)u^2.$

Un calcul pénible mais sans difficulté donne alors

$c_t(q!J(0,1)) = 1+(1-d)t+([\binom{d-1}{2}-g]t^2+gu)$

$-[2\binom{d-1}{3}-g(d-2)]tu+[2\binom{d-1}{4}+\binom{g}{2}-g(d-1)(d-4)/2]u^2$

$c_t(q!J(0,2)) = 1+(3-d)t+([\binom{d-2}{2}-g]t^2+[g+3-d]u)$

$-[2\binom{d-2}{3}-g(d-4)]tu+[(d-2)(d-3)^2(d-4)/12-g(d^2-7d+13-g)/2]u^2.$

Calculons enfin le genre arithmétique de la courbe T des trisécantes lorsque la variété des quadrisécantes est finie. Pour cela, considérons le θ_T-module sans torsion $M = R^1q_*J(0,1)$; son degré est c_4 $(q!J(0,1))$. En effet, M est un quotient de $q_*\theta_{\mathcal{C}}(0,1)$, donc est engendré par ses sections; si $\theta_T \to M$ est une section régulière de conoyau N, la présentation considérée de M induit la présentation suivante de N: $\theta_G \oplus R^1q_*E' \to R^1q_*E \to N \to 0$. On en déduit l'assertion par (2.1.II). Mais d'après (2.1.III), on a $\mathrm{Hom}_{\theta_T}(M,S_2(M)) \simeq M$ et $S_2(M) = \omega_T \otimes_{\theta_G} \theta_G(5-d)$, où ω_T est un 0_T-module dualisant; si g(T) est le genre arithmétique de T, ceci implique $d^oM = 2g(T)-2-(d-5)d^oT-d^oM$, soit $2g(T)-2 = 2c_4(q!J(0,1))+(d-5)d^oT$. On a donc démontré l'énoncé (bien connu) suivant:

Proposition 2.4: <u>Lorsque la variété des trisécantes est une courbe, son degré est</u> $[(d-1)(d-2)(d-3)/3-g(d-2)]$; <u>son genre arithmétique est</u>

$1+[(d-1)(d-2)(d-3)(3d-14)/12]+g(g-1)/2-g(d^2-6d+7)$

<u>lorsque la variété des quadrisécantes est finie. Dans ce cas, celle-ci est de degré</u>

$[(d-2)(d-3)^2(d-4)/12]-[g(d^2-7d+13-g)/2]$.

Supposons maintenant que les trisécantes forment une courbe T de G et que les quadrisécantes soient en nombre fini; nous savons que $M = R^1q_*J(0,1)$ est un θ_T-module sans torsion localement libre hors de la variété des quadrisécantes. La méthode de Kempf suggère d' éclater M (c'est à dire d'éclater un idéal fractionnaire isomorphe à M, où de facon équivalente de considérer $\mathrm{Proj}.(\oplus \bar{M}_j)$ où $\bar{M}_j$ est le quotient de $S_j(M)$ par son 0_T-torsion) pour résoudre les singularités de T (nous donnons plus loin une condition nécessaire et suffisante pour que cet éclatement soit non singulier lorsque C est une courbe lisse).

Proposition 2.5: <u>La courbe</u> T' <u>obtenue en éclatant</u> M <u>a genre arithmétique</u>

$[(d-4)(d-5)(2d-3)/6]+[(g/2)(d^2-9d+24-2g)]$.

Le genre arithmétique proposé pour T' est celui de T

diminué de trois fois le degré de la variété des quadrisécantes. Compte tenu de (I.3) c'est donc une conséquence de l'enoncé local suivant dans le cas $\ell-n = 2$.

Proposition 2.6: Soient R un anneau local régulier de dimension r, et $R^{\ell} \xrightarrow{\phi} R^{n}$ une application linéaire de matrice N, avec $r = \ell-n+2$ et $\ell \geq n$. Si $\mathfrak{a}_i$ est le i-ème idéal de Fitting de ϕ, supposons
(i) $\dim(R/\mathfrak{a}_0) = 1$, (ii) $\dim(R/\mathfrak{a}_1) = 0$ et
(iii) $\mathfrak{a}_1$ est engendré par les $(n-1)$-mineurs de N tronquée de sa première ligne. Alors, si $A = R/\mathfrak{a}_0$, et B est l'éclatement de $M = \operatorname{coker} \phi$, on a longueur $(B/A) = (\ell-n+1)$. longueur $(R/\mathfrak{a}_1)$.

Remarque: Pour $\ell = n$, on retrouve l'égalité de Gorenstein. Si L et L' sont des A-modules sans torsion de rang un, nous noterons $L.L'$ le quotient sans torsion de rang un de $L \otimes_A L'$ et M_i un tel produit de i facteurs égaux à M.

Soit x l'élément de M correspondant à la première ligne de la matrice N. Il induit pour $i \geq 0$ une application $M_i \xrightarrow{x} M_{i+1}$; montrons que longueur $(M_{i+1}/x.M_i)$ ne dépend pas de i pour $i \in [0,\ell-n]$. Si N' est la matrice obtenue en tronquant N de sa première ligne, on a $M_{i+1}/x.M_i = \operatorname{sym}_{i+1}(\operatorname{coker} N')$ pour ces valeurs de i et l'assertion résulte de (2.1(I)). Rappelons alors que $M_{\ell-n}$ est un A-module dualisant, et dualisons par rapport à ce module la suite exacte

$0 \longrightarrow M_{\ell-n} \xrightarrow{x} M_{\ell-n+1} \longrightarrow M_{\ell-n+1}/x.M_{\ell-n}$; on obtient alors

(2.1.III) une suite exacte
$0 \longrightarrow M_{-1} \xrightarrow{x} A \longrightarrow \operatorname{Ext}^1_A(M_{\ell-n+1}/x.M_{\ell-n}, M_{\ell-n}) \longrightarrow 0$, où

$M_{-1} \xrightarrow{x} A$ est l'application composée

$M_{-1} \longrightarrow M_{-1}.Ax \longrightarrow M_{-1}.M \longrightarrow A$ (cette dernière application étant décrite en (2.1.IV)). Mais le choix de x, et l'hypothèse faite sur les $(n-1)$-mineurs de N' entrainent que l'image de cette application est $\mathfrak{a}_1 A$. Il en résulte d'une part que $M_{-1}.Ax \to M_{-1}.M$ est un isomorphisme, et d'autre part que longueur $(A/\mathfrak{a}_1 A)$ = longueur $(M_{\ell-n+1}/x.M_{\ell-n})$. On a alors pour tout $k>0$ un isomorphisme $M_{-1}.Ax^k \simeq M_{-1}.M_k$.

Comme $B.M_k = M_k$ pour $k \gg 0$, on en déduit $B.M_{-1} = M_{-1}$ et $M_{-1} \otimes_B (B.M/Bx) = 0$, donc $B.M = Bx$ et $B.M_i = Bx^i$ pour $i \geq 0$. Mais par dualité (par rapport à $M_{\ell-n}$), $BM_{-1} = M_{-1}$ entraîne $B.M_{\ell-n+1} = M_{\ell-n+1}$ et finalement

$$\text{longueur}(B/A) = \text{longueur}(Bx^{\ell-n+1}/Ax^{\ell-n+1})$$

$$= \text{longueur}(M_{\ell-n+1}/Ax^{\ell-n+1}) = (\ell-n+1)\ \text{longueur}(A/\mathfrak{a}_1 A).$$

Proposition 2.7: <u>Si la courbe C est localement intersection complète les éclatements de $M = R^1q_*J(0,1)$ et de la variété des quadrisécantes dans la courbe des trisécantes sont isomorphes.</u>

Il est clair que la proposition est locale dans la grassmannienne. Soit donc R l'anneau local de G en une n-sécante ($n \geq 4$). L'anneau semi-local A de $\mathcal{C}$ au dessus de R est de la forme $R[Z]/(S,T)$ où S est un polynôme unitaire de degré n et T un polynôme de degré $\leq n-1$ (à coefficients dans l'idéal maximal de R). Notons $V = R[Z]/(S)$, c'est un R-module libre; rappelons que $\text{Hom}_R(V,R)$ est un V-module libre de rang un engendré par τ où $\tau(z^j) = \delta_{j,n-1}$ (ici z est l'image de Z dans V) pour $0 \leq j \leq n-1$. On vérifie directement que si $u_i = (T.\tau)(z^i)$ pour $i \geq 0$, alors $(u_{i+j})_{0\leq i,j\leq n-1}$ est la matrice de la multiplication par T dans le R-module libre V muni des bases $(1,z,\ldots,z^{n-1})$ et $(Q_0,\ldots,Q_{n-1})$ où $(z^j)^v = Q_j.\tau$. La courbe des trisécantes est donc définie dans R par le 2-ième idéal de Fitting de cette matrice; mais compte tenu des relations $u_{k+n} = \sum_{i=1}^n \alpha_i u_{k+n-i}$ où $S(Z) = Z^n - \sum_{i=1}^n \alpha_i Z^{n-i}$ (conséquences de la définition des u_i), elle est aussi définie par le 2-ième idéal de Fitting de la matrice infinie $(u_{i+j})_{\substack{0\leq i\leq n-1 \\ j\geq 0}}$. Mais d'aprés (1.4), cet idéal est aussi le 0-ième idéal de Fitting de la matrice $(u_{i+j})_{\substack{0\leq i\leq n-3 \\ j\geq 0}}$ et finalement de la matrice $(u_{i+j})_{\substack{0\leq i\leq n-3 \\ 0\leq j\leq n-1}}$ compte tenu des relations entre les u_i. On conclut alors par(2.2) qui donne ici un isomorphisme $M_4 \simeq \mathfrak{a}_1 A$ où A est l'anneau local de la courbe des trisécantes au point considéré de G, où $\mathfrak{a}_1$ est l'idéal de la variété des quadrisécantes dans R et où enfin M_4 est le quotient de $\mathrm{sym}_4(M)$ par son plus grand sous-module de longueur finie.

§ 3 Singularités des variétes de sécantes

On supposera toujours que C est une courbe lisse. L'anneau R sera le complété de l'anneau local de la grassmannienne au point L. Soit A^3 le complémentaire dans P^3 d'un plan ne contenant aucun des points P_i $(i=1,\dots,m)$ d'intersection de C et L. Choisissons un système de coordonnées (X,Y,Z) de A^3 tel que la droite $\overline{0Z}$ soit L et tel que le plan $Z = 0$ ne soit parallèle à aucune des tangentes à C aux points P_i. L'image réciproque de la variété d'incidence F sur $\mathrm{Spec}\, R[X,Y,Z]$ est définie par les équations $X = a+bZ$ et $Y = c+dZ$ où (a,b,c,d) est un système régulier de paramètres de R; nous l'identifierons donc à $\mathrm{Spec}\, R[Z]$. Si z_i est l'abscisse de P_i sur la droite $\overline{0Z}$, désignons par $A_i = R[[Z-z_i]]/\mathfrak{a}_i$ le complété de l'anneau local de $\mathcal{C}$ au point (L,P_i). Si n_i est la multiplicité de l'anneau A_i pour l'idéal maximal $\underline{m}$ de R, la droite L est une n-sécante de C avec $n = \sum_{i=1}^{m} n_i$. Décrivons maintenant l'idéal $\mathfrak{a}_i$ de A_i dans $R[[Z-z_i]]$, et pour cela considérons la représentation paramétrique de C au voisinage de P_i suivant l'uniformisante $(Z-z_i)$: $x = \phi_i(Z-z_i) = \sum_{j\geq n_i} \alpha_{i,j}(Z-z_i)^j$ et $y = \psi_i(Z-z_i) = \sum_{j\geq n_i} \beta_{ij}(Z-z_i)^j$ où on peut supposer $\alpha_{i,n_i} \neq 0$. L'idéal $\mathfrak{a}_i$ est donc engendré par les éléments $f_i = a+bZ-\phi_i$ et $g_i = c+dZ-\psi_i$. D'après le théorème de préparation de Weierstrass, il existe un polynôme distingué S_i de degré n_i en $(Z-z_i)$ engendrant le même idéal que f_i dans $R[[Z-z_i]]$ et un polynôme T_i de degré $\leq n_i-1$ en $(Z-z_i)$, à coefficients dans $\underline{m}$, tel que $T_i \equiv g_i \mod. S_i$.

Rappelons que la variété des r-sécantes a pour idéal dans R l'idéal de Fitting

$F^{r-1}(\oplus_1^m A_i) = \sum_{j_1+\ldots+j_m=r-1} F^{j_1}(A_1)\times\ldots\times F^{j_m}(A_m)$; en particulier l'idéal de la variété des n-sécantes est $\sum_1^m F^{n_i-1}(A_i)$, car pour $i = 1,\ldots,m$, le R-module A_i est engendré par n_i éléments. Il nous semble convenable d'étendre ici notre définition globale des variétés de sécantes dans le cadre local de la manière naturelle qui suit:

Définition: <u>L'idéal dans</u> R <u>de la variété des</u> r-<u>sécantes aux voisinages des points</u> $P_{i_1},\ldots,P_{i_k}$ <u>de</u> $L \cap C$ <u>est</u> $F^{r-1}(\oplus_{j=1}^k A_{i_j})$.

Commençons par étudier la variété des n_i-sécantes au voisinage du point P_i, dont l'idéal dans R est $F^{n_i-1}(A_i)$. On démontre naturellement que cet idéal est engendré par les coefficients du polynôme T_i. Calculons les termes de degré un du développement en série formelle (en a,b,c,d) de ces coefficients; ils sont identiques aux termes de degré un (en a,b,c,d) des n_i premiers coefficients de la série formelle (en $(Z-z_i)$)
$g_i - f_i[(\sum_{n\geq n_i}\beta_{i,n}(Z-z_i)^{n-n_i})/(\sum_{n\geq n_i}\alpha_{i,n}(Z-z_i)^{n-n_i}]$. Ils engendrent donc le même idéal que les termes de degré un (en a,b,c,d) des n_i premiers coefficients de la série formelle $g_i[\sum_{n\geq n_i}\alpha_{i,n}(Z-z_i)^{n-n_i}] - f_i[\sum_{n\geq n_i}\beta_{i,n}(Z-z_i)^{n-n_i}]$; la matrice M_i de ces termes dans la base (a,b,c,d) est

$$\begin{pmatrix} -\beta_{i,n_i}, & -\beta_{i,n_i+1} & ,\ldots, & -\beta_{i,2n_i-1} \\ -\beta_{i,n_i}\cdot z_i, & -\beta_{i,n_i+1}\cdot z_i-\beta_{i,n_i} & ,\ldots, & -\beta_{i,2n_i-1}\cdot z_i-\beta_{i,2n_i-2} \\ \alpha_{i,n_i}, & \alpha_{i,n_i+1} & ,\ldots, & \alpha_{i,2n_i-1} \\ \alpha_{i,n_i}\cdot z_i, & \alpha_{i,n_i+1}\cdot z_i+\alpha_{i,n_i} & ,\ldots, & \alpha_{i,2n_i-1}\cdot z_i+\alpha_{i,2n_i-2} \end{pmatrix}$$

Il est alors clair que la codimension de l'espace tangent de Zariski à la variété des n'-secantes (avec $n' = \sum_{j=1}^k n_{i_j}$), aux voisinages des points $P_{i_1},\ldots,P_{i_k}$, est le rang de la matrice à 4 lignes et n' colonnes $M = (M_{i_1},\ldots,M_{i_k})$. On en déduit l'énoncé suivant qui en particulier décrit pour tout n le lieu singulier de la variété des n-sécantes, hors de la variété des (n+1)-sécantes.

Proposition 3.1: Soient L une droite et $P_1,\ldots,P_m$ des points distincts de $L \cap C$. Si $n_i = \mathrm{rg}(\theta_{L,P_i} \otimes_{\theta_{P^3}} \theta_{C,P_i})$, soit $n = \sum_1^m n_i$. Supposons $n \geq 2$.

(I) La dimension de l'espace tangent de Zariski à la variété des n-sécantes aux voisinages des points P_i $(i=1,\ldots,m)$, au point L de G est toujours ≤ 2.

(II) La dimension de cet espace tangent est ≥ 1 si et seulement si il existe une quadrique Q contenant L et non singulière en un point général de L, telle que la multiplicté d'intersection de Q et C en P_i est $\geq 2n_i$ pour $i = 1,\ldots,m$.

(III) Pour $n \geq 3$, la dimension de cet espace tangent est 2 si et seulement si il existe un plan H tel que la multiplicité d'intersection de H et C en P_i est $\geq 2n_i$ pour $i = 1,\ldots,m$.

Comme $n \geq 2$, la matrice $M = (M_1,\ldots,M_m)$ contient soit une matrice de la forme

$$\begin{pmatrix} -\beta_{11} & , -\beta_{21} \\ -\beta_{11}\cdot z_1 & , -\beta_{21}\cdot z_2 \\ \alpha_{11} & , \alpha_{21} \\ \alpha_{11}\cdot z_1 & , \alpha_{21}\cdot z_2 \end{pmatrix}, \text{ où } \alpha_{11}\neq 0,\ \alpha_{21}\neq 0,\ z_1\neq z_2,$$

soit une matrice de la forme

$$\begin{pmatrix} -\beta_{12} & , -\beta_{13} \\ -\beta_{12}\cdot z_1 & , -\beta_{13}\cdot z_1-\beta_{12} \\ \alpha_{12} & , \alpha_{13} \\ \alpha_{12}\cdot z_1 & , \alpha_{13}\cdot z_1+\alpha_{12} \end{pmatrix}, \text{ où } \alpha_{12}\neq 0.$$

Il est clair que ces matrices sont de rang deux, ce qui démontre (I). Dégageons ensuite le résultat suivant:

Lemme 3.2: $(\gamma,\delta,\eta,\lambda)$ <u>est une relation entre les lignes de la matrice</u> M_i <u>si et seulement si la surface d'équation affine</u> $\lambda XZ+\eta X+\delta YZ+\gamma Y = 0$ <u>a une multiplicité d'intersection</u> $\geq 2n_i$ <u>avec</u> C <u>en</u> P_i .

Considérons la représentation paramétrique de C au voisinage de P_i: $x = \sum_{j\geq n_i} \alpha_{ij}(Z-z_i)^j$; $y = \sum_{j\geq n_i} \beta_{ij}(Z-z_i)^j$; $z = z_i+(Z-z_i)$. Remplaçons x,y et z par ces expressions dans $\lambda xz+\eta x+\delta yz+\gamma z$; on vérifie directement que $(\gamma,\delta,\eta,\lambda)$ est une relation entre les lignes de M_i si et seulement si la série formelle obtenue est d'ordre $\geq 2n_i$ en $(Z-z_i)$.

Pour montrer II, supposons d'abord $\operatorname{rg} M \leq 3$; soit alors $(\gamma,\delta,\eta,\lambda)$ une relation non nulle entre les lignes de M. La quadrique (éventuellement dégénérée) d'équation affine $\lambda XZ+\eta X+\delta YZ+\gamma Y = 0$ ne contenant pas deux fois la droite $L = \overline{0Z}$, le lemme exprime qu'elle a les propriétés d'intersection demandées. Réciproquement, soit $\lambda XZ+\eta X+\delta YZ+\gamma Y+\varepsilon X^2+\mu Y^2+\nu XY = 0$ l'équation affine de la quadrique Q; comme la quadrique d'équation $\varepsilon Z^2+\mu Y^2+\nu XY = 0$ a évidemment les propriétés d'intersection requises pour Q, celle dont l'équation est $\lambda XZ+\eta X+\delta YZ+\gamma Y = 0$ aussi. Mais $(\lambda,\eta,\delta,\gamma) \neq 0$ car Q ne contient pas deux fois L, donc c'est une relation non triviale entre les lignes de M.

Pour démontrer III, supposons d'abord l'espace tangent de dimension deux et considérons le pinceau de quadriques ainsi défini. Pour $n \geq 3$, il contient un plan fixe; sinon, soit Γ la courbe de degré trois résiduelle de L dans le pinceau. On vérifie facilement que $L \cap \Gamma$ est une variété finie de degré deux; d'autre part, il est clair que $L \cap \Gamma$ contient le point P_i pris n_i fois sur L. Ceci contredit $n \geq 3$. Soit donc $\eta X+\gamma Y = 0$ l'équation de ce plan fixe; une quadrique du pinceau a alors une équation affine de la forme $(\eta X+\lambda Y)(\mu Z+\nu) = 0$. Mais pour (μ,ν) suffisamment général, le plan d'équation $\mu Z+\nu = 0$ ne passe par aucun des points P_i , donc le plan fixe a la propriété requise. Réciproquement, si H est un plan ayant cette propriété, son équation affine est nécessairement de la forme

$\eta X+\lambda Y=0$; il est alors clair que $(\eta,0,\lambda,0)$ et $(0,\eta,0,\lambda)$ sont deux relations indépendantes entre les lignes de M.

La fin de ce paragraphe est consacrée à l'étude des singularités de la courbe obtenue par éclatement, dans la courbe des trisécantes, de la variété des quadrisécantes, lorsque celle-ci est finie. Le problème étant local, nous conservons les notations utilisées plus haut et identifions maintenant l'anneau de $\mathcal{C}$ au-dessus du complété R de l'anneau local du point considéré L de G à $R[Z]/(S,T)$ où S est un polymôme unitaire de degré n (la droite L est une n-sécante) et T un polynôme de degré $\leq n-1$ à coefficients dans l'idéal maximal $\underline{m}$ de R. Toujours avec les mêmes notations, la réduction de S modulo $\underline{m}$ est $\bar{S} = \Pi_{i=1}^{m}(Z-z_i)^{n_i}$. Nous utiliserons aussi la présentation persymétrique de l'anneau de la courbe, consideré comme R-module, décrite dans la démonstration de (2.7): si τ est la R-forme linéaire sur $R[Z]/(S)$ telle que $\tau(\bar{Z}^j) = \delta_{n-1,j}$ pour $0 \leq j \leq n-1$ (où $\bar{Z}$ designe l'image de Z dans $R[Z]/(S)$), et si $u_i = \tau(T.\bar{Z}^i)$ pour $i \geq 0$, alors $(u_{i+j})_{0\leq i,j\leq n-1}$ est la matrice de la multiplication par T dans $R[Z]/(S)$ muni des bases $\bar{Z}^j$ $(0\leq j\leq n-1)$ et Q_j où $(\bar{Z}^j)^v = Q_j.\tau$. Rappelons que si $S = s_0+s_1Z+\ldots+s_{n-1}Z^{n-1}+Z^n$, les éléments u_j vérifient les relations
$s_0u_k+s_1u_{k+1}+\ldots+s_{n-1}u_{k+n-1}+u_{k+n} = 0$ pour $k \geq 0$.

Si A est le complété de l'anneau local de la courbe des trisécantes au point L, son idéal I dans R est engendré par les mineurs maximaux de la matrice $(u_{i+j})_{\substack{0\leq i\leq n-3\\ 0\leq j\leq n-1}}$; de plus les mineurs sous-maximaux de cette matrice engendrent l'idéal $\mathfrak{a}$ de la variété des quadrisécantes dans R (donc un idéal de définition de R).

Proposition 3.3: <u>L'anneau de l'éclatement de</u> A <u>dans</u> A <u>est isomorphe</u> à $R[X_0,\ldots,X_{n-4}]/(V_0,V_1,V_2,r_0,\ldots,r_{n-4})$ <u>où</u> X_i $(0\leq i\leq n-4)$ <u>sont des indéterminées</u>, $V_i = u_iX_0+u_{i+1}X_1+\ldots+u_{i+n-4}X_{n-4}+u_{i+n-3}$ <u>pour</u> $i \geq 0$ <u>et</u> $r_0+r_1Z+\ldots+r_{n-4}Z^{n-4}$ <u>est le reste de la division euclidienne de</u> S <u>par</u> $X_0+X_1Z+\ldots+X_{n-4}Z^{n-4}+Z^{n-3}$ <u>dans</u> $R[X_0,\ldots,X_{n-4}][Z]$.

Soit M le conoyau de l'application linéaire $R^n \to R^{n-2}$ définie par la matrice $(u_{i+j})_{\substack{0\leq i\leq n-3\\ 0\leq j\leq n-1}}$. Nous avons vu en (2.7) que l'éclatement de $\mathfrak{a}A$ dans A est isomorphe à l'éclatement de M, donc à Proj. $(\oplus_{i\geq 0}\bar{M}_i)$ où $\bar{M}_i$ est le A-module sans torsion de rang un quotient de $\text{sym}_i(M)$. Notons que l'algèbre symétrique de M est isomorphe à $R[Y_0,\dots,Y_{n-3}]/(U_0,\dots,U_{n-1})$ où $U_j = u_jY_0+\dots+u_{j+n-3}Y_{n-3}$ pour $j \geq 0$, et remarquons de plus que compte tenu des relations entre les u_i, il est clair que $U_j\in(U_0,\dots,U_{n-1})$ pour $j \geq 0$. Considérons le plongement de l'éclatement E dans P_R^{n-3} ainsi décrit. Montrons d'abord que l'hyperplan $Y_{n-3} = 0$ coupe proprement E; en effet, si q est un idéal premier gradué de $R[Y_0,\dots,Y_{n-3}]$ contenant Y_{n-3} et U_j $(j\geq 0)$, on remarque que $(Y_0,\dots,Y_{n-4},0)$ et $(0,Y_0,\dots,Y_{n-4})$ sont deux relations, modulo q, indépendantes entre les lignes de la matrice $(u_{i+j})_{\substack{0\leq i\leq n-3\\ 0\leq j\leq n-1}}$ donc q contient $\mathfrak{a}$ et finalement $\underline{m}$. Plaçons nous maintenant dans l'ouvert complémentaire de l'hyperplan $Y_{n-3} = 0$; notons $X_i = Y_i/Y_{n-3}$ et $V_i = U_i/Y_{n-3}$. Soit $B'=R[X_0,\dots,X_{n-3}]/(V_0,V_1,V_2,r_0,\dots,r_{n-4})$.

Désignons par Ω le polymôme $X_0+X_1Z+\dots+X_{n-4}Z^{n-4}+Z^{n-3}$, et soit $\Delta = \delta_0+\delta_1Z+\delta_2Z^2+Z^3$ le polynome à coefficients dans $R[X_0,\dots,X_{n-4}]$ tel que $S = \Omega.\Delta+\sum_{i=0}^{n-4} r_iZ^i$. On a évidemment $S_{B'} = \Omega_{B'}.\Delta_{B'}$, où $S_{B'},\Omega_{B'},\dots$ sont les réductions à $B'[Z]$ des polymômes considérés. On en déduit que B' est entier sur R puisque S ayant ses coefficients dans R, les coefficients de $\Omega_{B'}$ sont entiers sur R. Remarquons ensuite que $V_k = \tau(\bar{Z}^k.T.\Omega)$ où la R-forme linéaire τ sur $R[Z]/(S)$ est naturellement étendue en une $R[Z_0,\dots,X_{n-4}]$-forme linéaire sur $R[X_0,\dots,X_{n-4}][Z]/(S)$. On en déduit

$\delta_0V_k+\delta_1V_{k+1}+\delta_2V_{k+2}+V_{k+3} = \tau(\bar{Z}^k.T.\Omega.\Delta)$; mais comme l'image de $\Omega.\Delta$ dans $B'[Z]/(S_{B'})$ est nulle, ceci implique que l'image de $\delta_0V_k+\delta_1V_{k+1}+\delta_2V_{k+2}+V_{k+3}$ dans B' est nulle pour $k\geq 0$, et finalement que $V_k \in (V_0,V_1,V_2,r_0,\dots,r_{n-4})$ pour $k\geq 0$. Il en résulte que B' est une A-algèbre finie. Mais alors tout idéal maximal de B' contient $\underline{m}A$, donc est l'image d'un idéal maximal de hauteur n de $R[X_0,\ldots,X_{n-4}]$; comme l'idéal de B' dans cet anneau est engendré par (n-1) éléments, il est clair que B' est une intersection complète. Pour conclure la démonstration, il suffit de vérifier que les images des r_i dans l'anneau B de E dans l'ouvert affine $Y_{n-3} \neq 0$ sont nulles. En effet, B sera alors le quotient de B' par son plus grand idéal de longueur finie, donc on aura B = B' (puisque B' n'a pas d'idéal de longueur finie); de plus il y a au plus un nombre fini de points de E dans l'hyperplan $Y_{n-3} = 0$, mais B entier sur A entraîne alors E = Spec.B . Remarquons alors que la multiplication par $T_B.\Omega_B$ est nulle dans $B[Z]/(S_B)$ puisque l'image de V_j dans B est nulle pour $j > 0$; on en déduit que $T_B.\Omega_B$ est divisible par S_B dans $B[Z]$, donc $T_B.(\sum_0^{n-4} r_i Z^i)$ aussi. Mais la matrice de la multiplication par T_B dans les bases images de (Z^i) et (Ω_i) $(0 \leq i \leq n-1)$ dans $B[Z]/S_B)$ est $(\bar{u}_{i+j})_{0 \leq i,j \leq n-1}$ (où $\bar{u}_k$ est l' image de u_k dans B). Comme les mineurs maximaux de $(\bar{u}_{i+j})_{\substack{0 \leq i \leq n-4 \\ 0 \leq j \leq n-1}}$ engendrent $\mathfrak{a}B$ (d'apres 2.3 et les relations entre les u_i , par exemple), et comme cet idéal est sans B-torsion, la restriction de la multiplication par T_B au sous B-module libre de $B[Z]/(S_B)$ engendré par $\bar{Z}^i$ $(0 \leq i \leq n-4)$ est injective; donc $(\sum_0^{n-4} r_i Z^i)_B$ est divisible par S_B , ce qui entraîne évidemment $(\sum_0^{n-4} r_i Z^i)_B = 0$.

La proposition étant établie, notons que les idéaux maximaux de B correspondent aux points $x = (x_0,\ldots,x_{n-4})$ de k^{n-3} tels que $\Omega_x = x_0 + x_1 Z + \ldots + x_{n-4} Z^{n-4} + Z^{n-3}$ divise $\bar{S}$. Fixons un tel point, et notons $\Delta_x = \delta_0(x) + \delta_1(x) Z + \delta_2(x) Z^2 + Z^3$ la réduction de Δ à ce point. On a alors $\Delta_x = \Pi_{i=1}^m (Z - z_i)^{n_{i,x}}$ avec $\sum_{i=1}^m n_{i,x} = 3$, et $\Omega_x = \Pi_{i=1}^m (Z - z_i)^{n_i - n_{i,x}}$.

Théorème 3.4: <u>Pour que</u> E <u>soit non singulier au point</u> x, <u>il faut et il suffit qu'il n'existe pas de plan</u> H <u>de</u> P^3 <u>dont la multiplicité d'intersection avec</u> C <u>en</u> P_i <u>soit</u> $\geq 2n_{i,x}$ <u>pour</u> $i = 1,\ldots,m$.

Posons $Y_i = (x_i - X_i)$ pour $i = 0,\ldots,n-4$, et considérons l'idéal maximal $\underline{n} = (\underline{m}, Y_0,\ldots,Y_{n-4})$ de $R[X_0,\ldots,X_{n-4}]$. Pour que E soit non singulier en x, il faut et il suffit que les classes de $V_0, V_1, V_2, r_0,\ldots,r_{n-4}$ dans $\underline{n}/\underline{n}^2$ engendrent un sous-espace vectoriel de codimension un. Soit N le conoyau de l'application injective naturelle $\underline{m}/\underline{m}^2 \to \underline{n}/\underline{n}^2$. Il est clair que les classes des éléments V_0, V_1, V_2 sont dans $\underline{m}/\underline{m}^2$; notons M le sous-espace vectoriel de $\underline{n}/\underline{n}^2$ engendré par les classes de $r_0,\ldots,r_{n-4}$.

Lemme 3.5: <u>L'image de l'application</u> $M \to N$ <u>est de codimension égale au degré du p.g.c.d. de</u> Ω_x <u>et</u> Δ_x. <u>L'intersection de</u> M <u>et</u> $\underline{m}/\underline{m}^2$ <u>est engendrée par les classes des coefficients du reste de la division euclidienne de</u> S <u>par le p.g.c.d. de</u> Ω_x <u>et</u> Δ_x.

Notons $\sum_{i=0}^{n-4} k_i Z^i$ (resp. $\sum_{i=0}^{n-4} \ell_i Z^i$) le reste de la division euclidienne de S (resp. $(\sum_0^{n-4} Y_i Z^i).\Delta_x$) par Ω_x. Utilisant l'identité $\Omega = \Omega_x - \sum_0^{n-4} Y_i Z^i$, on démontre facilement $r_i \equiv k_i + \ell_i \mod.\underline{n}^2$. L'image de $M \to N$ est alors engendrée par les images des classes $\bar{\ell}_i$ dans $\underline{N}/\underline{N}^2$ des éléments ℓ_i, et le noyau de cette application est l'espace vectoriel $M \cap (\underline{m}/\underline{m}^2)$ des combinaisons linéaires, à coefficients dans le corps de base, $\sum_0^{n-4} c_i \bar{k}_i$ (où $\bar{k}_i$ est la classe de k_i dans $\underline{m}/\underline{m}^2$) telles que $\sum_0^{n-4} c_i \bar{\ell}_i = 0$. Identifions alors $N^\vee$ et $k[Z]/\Omega_x$ en identifiant la base duale de la base des images des classes des Y_i et la base des classes des Z^i pour $0 \leqslant i \leqslant n-4$. Pour tout $f \in N^\vee$, la congruence $(\sum_0^{n-4} Y_i Z^i).\Delta_x \equiv \sum_0^{n-4} \ell_i Z^i \mod.\Omega_x$ donne une nouvelle congruence $(\sum_0^{n-4} f(\bar{Y}_i) Z^i).\Delta_x \equiv \sum_0^{n-4} f(\bar{\ell}_i) Z^i \mod.\Omega_x$ dans $k[Z]$. Ceci montre bien que l'orthogonal de l'image de M dans N est de rang égal au degré du p.g.c.d. de Ω_x et Δ_x. D'autre part, la relation $\sum_0^{n-4} c_i \bar{\ell}_i = 0$ est équivalente à $\sum_0^{n-4} f(\bar{\ell}_i) c_i = 0$ pour tout $f \in N^\vee$; mais si π est la forme linéaire sur $R[Z]/\Omega_x$ telle que $\pi(Z^i) = c_i$ pour $0 \leqslant i \leqslant n-4$, cette dernière propriété est équivalente à $\pi([\sum_0^{n-4} f(\bar{Y}_i) Z^i].\Delta_x) = 0$ pour tout $f \in N$, autrement dit π

est orthogonale à l'idéal de $k[Z]/\Omega_x$ engendré par Δ_x. Finalement, les combinaisons linéaires $\sum_0^{n-4} c_i\bar{k}_i$ telles que $\sum_0^{n-4} c_i\bar{\ell}_i = 0$ sont les classes dans $\underline{m}/\underline{m}^2$ des images de S par les formes R-linéaires sur $R[Z]/\Omega_x$ s'annulant sur l'idéal engendré par Δ_x; on vérifie facilement que ces images forment l'idéal des coefficients du reste de la division euclidienne de S par le p.g.c.d. de Ω_x et Δ_x. Le lemme étant démontré, nous sommes amenés à distinguer trois cas:

1) Lorsque Ω_x et Δ_x sont premiers entre eux, l'application $M \to N$ est un isomorphisme; dans ce cas, x est un point non singulier si et seulement si les classes de V_0, V_1, V_2 engendrent un sous-espace vectoriel de rang trois de $\underline{m}/\underline{m}^2$. Remarquons que dans ce cas la décomposition $\bar{S} = \Omega_x.\Delta_x$ se relève en une décomposition $S = S'.S''$ dans $R[Z]$, où S'' est le polynôme unitaire de degré trois associé aux branches des points P_i pour lesquels $n_{i,x} \neq 0$, donc $n_{i,x} = n_i$. Les classes des éléments $x_0u_i + x_1u_{i+1} + \ldots, + x_{n-4}u_{i+n-4} + u_{i+n-3}$ $(i=0,1,2)$ engendrent alors le même sous-espace vectoriel de $\underline{m}/\underline{m}^2$ que les classes des coefficients du reste de la division euclidienne de T par S''. Nous avons vu plus haut (3.1) que ce sous-espace est de rang trois si et seulement s'il n'existe pas de plan H dont la multiplicité d'intersection avec C en P_i est $\geq 2n_{i,x}$.

2) Lorsque Ω_x et Δ_x ont un facteur commun de degré deux, il ressort de (3.5) que E est singulier en x. Soit $\Pi_1^m(Z-z_i)^{n'_{i,x}}$ un facteur commun de degré deux de Ω_x et Δ_x, et soit P_j le point tel que $\Delta_x = (Z-z_j)\Pi_1^m(Z-z_i)^{n'_{i,x}}$. Alors $n_j \geq 1+2n'_{j,x}$; il est clair qu'il existe un plan H contenant L dont la multiplicité d'intersection avec C en P_j est $\geq 2+2n'_{j,x} = 2n_{j,x}$, et comme la multiplicité d'intersection de H et C en P_i $(i \neq j)$ est nécessairement $\geq 2n'_{i,x}$, le plan H a les propriétés d'intersection requises car $n_{i,x} = n'_{i,x}$ pour $i \neq j$.

3) Lorsque Ω_x et Δ_x ont un p.g.c.d. de degré un, on peut supposer que ce p.g.c.d. est Z et correspond au

point P_1 (i.e. $z_1=0$). D'après (3.5), pour que E soit non singulier en x il faut et il suffit que les classes de V_0, V_1, V_2 et $S(0)$ engendrent $\underline{m}/\underline{m}^2$. On vérifie d'abord que les classes de (V_0, V_1, V_2) engendrent le même sous-espace vectoriel de $\underline{m}/\underline{m}^2$ que les classes des coefficients du reste de la division euclidienne de T par Δ_x; ce sous espace vectoriel est aussi engendré par les classes des coefficients des restes des divisions euclidiennes de T par les polymômes $(Z-z_i)^{n_{i,x}}$ pour $n_{i,x} \neq 0$. Utilisons alors le système régulier de paramètres (a,b,c,d) défini au début de ce paragraphe, et remarquons que compte tenu de la construction de S, les classes de $S(0)$ et a sont les mêmes à un coefficient inversible près. Traitons séparément les quatre cas qui suivent:

(i) Si $\Delta_x = Z^3$, donc $\Omega_x = Z.\Pi_2^m(Z-z_i)^{n_i}$; pour que E soit non singulier en x il faut et il suffit que

$$\begin{vmatrix} 0 & -\beta_{14} & -\beta_{15} \\ \alpha_{14} & \alpha_{15} & \alpha_{16} \\ 0 & \alpha_{14} & \alpha_{15} \end{vmatrix} \neq 0 \text{ , soit } \begin{vmatrix} \beta_{14} & \beta_{15} \\ \alpha_{14} & \alpha_{15} \end{vmatrix} \neq 0 ,$$

c'est-à-dire qu'il n'existe pas de plan dont la multiplicité d'intersection avec C en P_1 soit > 6.

(ii) Si $\Delta_x = Z^2(Z-z_2)$, donc $\Omega_x = Z.\Pi_3^m(Z-z_i)^{n_i}$; la condition de non singularité en x est

$$\begin{vmatrix} 0 & -\beta_{13} & -\beta_{21}\cdot z_2 \\ \alpha_{13} & \alpha_{14} & \alpha_{21} \\ 0 & \alpha_{13} & \alpha_{21}\cdot z_2 \end{vmatrix} \neq 0 \text{ , soit } \begin{vmatrix} \beta_{13} & \beta_{21} \\ \alpha_{13} & \alpha_{21} \end{vmatrix} \neq 0 ,$$

c'est-à-dire qu'il n'existe pas de plan dont la multiplicité d'intersection avec C soit > 4 en P_1, et > 2 en P_2.

(iii) Si $\Delta_x = Z(Z-z_2)^2$, donc $\Omega_x = Z^{n_1-1}.\Pi_3^m(Z-z_i)^{n_i}$, la condition de non singularité en x est

$$\begin{vmatrix} 0 & -\beta_{22}\cdot z_2 & -\beta_{23}\cdot z_2-\beta_{22} \\ \alpha_{1n_1} & \alpha_{22} & \alpha_{23} \\ 0 & \alpha_{22}\cdot z_2 & \alpha_{23}\cdot z_2+\alpha_{22} \end{vmatrix} \neq 0 \text{ , soit } \begin{vmatrix} \beta_{22} & \beta_{23} \\ \alpha_{22} & \alpha_{23} \end{vmatrix} \neq 0,$$

c'est-à-dire qu'il n'existe pas de plan dont la multiplicité d'intersection avec C soit ≥ 4 en P_2, et (nécessairement) ≥ 2 en P_1.

(iv) Si $\Delta_x = Z(Z-z_2)(Z-z_3)$, donc $\Omega_x = Z^{n_1-1}\ \Pi_4^m(Z-z_i)^{n_i}$, la condition de non singularité pour E en x devient

$$\begin{vmatrix} 0 & , -\beta_{21}\cdot z_2 & , -\beta_{31}\cdot z_3 \\ \alpha_{1n_1}, & \alpha_{21} & , \alpha_{31} \\ 0 & , \alpha_{21}\cdot z_2 & , \alpha_{31}\cdot z_3 \end{vmatrix} \neq 0 \text{ , soit } \begin{vmatrix} \beta_{21} & , \beta_{31} \\ \alpha_{21} & , \alpha_{31} \end{vmatrix} \neq 0,$$

c'est-à-dire qui il n'existe pas de plan tangent à C en P_2 et P_3 et ayant (nécessairement) une multiplicité d'intersection ≥ 2 avec C en P_1.

Compte tenu de (2.5) et de (3.1), nous avons aussi démontré le résultat suivant:

Théorème 3.6: Pour que la courbe des trisécantes d'une courbe lisse de degré d et genre g soit réduite de genre géométrique $(d-4)(d-5)(2d-3)/6 + g(d^2-9d+24-2g)/2$, il faut et il suffit que la courbe n'ait qu'un nombre fini de n-sécantes ($n\geq 4$) et qu'il n'existe pas un plan H, une droite L de H et des points P_i $(i=1,\ldots,m)$ de $L \cap C$ vérifiant la condition suivante:
Il existe des entiers positifs n_i $(i=1,\ldots,m)$, avec $\sum_1^m n_i = 3$, tels que $rg(\theta_{L,P_i} \otimes_{O_{P^3}} \theta_{C,P_i}) \geq n_i$ et que la multiplicité d'intersection de H et C en P_i est $\geq 2n_i$, pour $i = 1,\ldots,m$.

Remarque: Il serait naturel de chercher à calculer la correction à apporter au genre géometrique pour une trisécante "singulière" au sens défini dans le théorème.

Remarque 3.7: On peut montrer que l'éclatement de la variété des quadrisécantes dans la courbe des trisécantes est naturellement plongé dans C^3 divisée par S_3. C'est le point de vue des auteurs modernes, et particulièrement de Schwarzenberger ([4]). Le calcul du genre arithmétique de cette courbe peut se faire dans ce cadre (il est simple dans le cas des courbes rationnelles, car alors $S_3(C) = P^3$).

Nous ne savons pas si cette approche permet aussi de déterminer les singularités de cette courbe.

Appendice A. Adjointes itérées.

Soit C une courbe réduite tracée sur une surface lisse S (schéma excellent de dimension 2), et soit C' sa normalisée.

Définition A.1: <u>Une courbe</u> Γ <u>de</u> S <u>sera dite adjointe d'ordre</u> i <u>de</u> C <u>si son idéal est contenu dans</u> $F^i(\theta_{C'})$, <u>le</u> i-<u>ième idéal de Fitting du</u> 0_S-<u>module</u> $\theta_{C'}$.

On a vu (1.5) que pour $i = 1$, on retrouve la définition classique. L'enoncé qui suit est bien connu pour les adjointes du premier ordre (voir par exemple [6]).

Proposition A.2: Γ <u>est une adjointe d'ordre</u> i <u>de</u> C <u>si et seulement si pour tout point</u> P, <u>infiniment voisin d'un point de</u> S, <u>on a</u> $e_P(\Gamma) \geqslant e_P(C)-i$ (<u>où</u> $e_P(\cdot)$ <u>désigne la multiplicité en</u> P).

Par une récurrence sur le nombre minimum d'éclatements de points nécessaire pour résoudre les singularités de C, la proposition se déduit du lemme suivant.

Lemme A.3: <u>Soient</u> R <u>un anneau local régulier de dimension deux, d'idéal maximal</u> $\underline{m}$, <u>et</u> A <u>un quotient réduit de dimension un, de multiplicité</u> e <u>pour</u> $\underline{m}$. <u>Soient</u> B <u>l'éclatement de</u> $\underline{m}A$ <u>dans</u> A, <u>et</u> R' <u>l'anneau semi--local de l'éclatement de</u> $\underline{m}$ <u>dans</u> R <u>aux points fermés de</u> Spec.B. . <u>Si</u> M <u>est un</u> B-<u>module de type fini, sans torsion, de rang un, on a</u> $R'.F^i_R(M) = \underline{m}^{e-i}.F^i_{R'}(M)$ (<u>où</u> l'<u>on a précisé l'anneau de base dans la notation de l'idéal de Fitting</u>).

Remarquons que $M/\underline{m}M$ est un $(R/\underline{m})$-espace vectoriel de rang e ; en effet, la multiplicité de M est e, donc le rang de $\underline{m}^nM/\underline{m}^{n+1}M$ est e pour $n \gg 0$, mais comme M est un B-module, on a $\underline{m}M \simeq M$. Condidérons alors (α_{ij}) une matrice $e{\times}e$, à coefficients dans $\underline{m}$, présentation du R-module M, correspondant au choix d'un système minimal de générateurs $(x_1,\dots,x_e)$ de M. Soit a un générateur de $\underline{m}R'$. Posons $\beta_{ij} = \alpha_{ij}\, a^{-1}$; ce sont des éléments de R'. Il suffit de montrer que (β_{ij}) est une présentation du R'-module M. Si $f = |\alpha_{ij}|$, c'est une équation de A dans R, et $g = f.a^{-e} = |\beta_{ij}|$ est une équation de B dans R'. Soit (γ_{ij}) une présentation du R'-module M correspondant au système de générateurs $(x_1,\dots,x_e)$. Comme a est régulier dans M, (β_{ij}) est une matrice de relations (à coefficients dans R') entre les éléments $x_1,\dots,x_e$ de M. Les matrices (β_{ij}) et (γ_{ij}) ayant même déterminant (à un coefficient inversible près), elles sont équivalentes, ce qui démontre le lemme.

Proposition A.4: Si J_i est l'idéal dans P^2 des adjointes d'ordre i d'une courbe intègre C, de degré d, de P^2, alors $H^1(J_i(n)) = 0$ pour $n \geq d-2i-1$ si et seulement si C n'a pas de point multiple d'odre $\geq d-i+1$.

Montrons que la condition est suffisante, la réciproque se vérifiant sans difficultés. Le résultat est classique pour $i = 1$. Supposons donc $i \geq 2$. Remarquons que d'après (A.2) une adjointe d'ordre r d'une adjointe d'ordre s de C est une adjointe d'ordre $r+s$ de C. Par une récurrence évidente sur i, il suffit alors de montrer qu'une première adjointe générale de C, de degré $d-2$, est intégre et n'a pas de point multiple d'ordre $d-i = (d-2)-(i-1)+1$. Cette dernière propriété se déduit immédiatement du fait que $J_1(d-2)$ est engendré par ses sections, ce qui est une conséquence facile de la régularité de l'adjointe. Pour montrer qu'une section générale de $J_1(d-2)$ est intègre, considérons le morphisme de l'éclatement $\bar{P}$ de J_1 dans $P^{g+d-2} = P(H^o(J_1(d-2)))$, où g est le genre géométrique de C. Si son image est une surface, on a terminé d'aprés le théorème de Bertini. Sinon, c'est une courbe Γ de P^{g+d-2}; comme Γ est gauche, son degré δ est $\geq g+d-2$.

Une section hyperplane du morphisme $\bar{P} \to P^{g+d-2}$ est la transformée propre dans $\bar{P}$ d'une première adjointe de degré d-2; comme elle a, en général, δ composantes irréductibles, on en déduit $g = 0$ et $\delta = d-2$, donc $\Gamma = P^1$. Soit L le faisceau inversible sur $\bar{P}$ image réciproque de $O_{P^1}(1)$; il est clair qu'une section de L est la transformée propre d'une droite de P^2. Ceci induit une application linéaire non nulle de L dans $O_{\bar{P}}(1)$, l'image réciproque de $O_{P^2}(1)$. L'image directe de L sur P^2 est donc un sous faisceau de $O_{P^2}(1)$ ayant deux sections. Ces deux sections définissent un point Q de P^2, les adjointes de degré d-2 sont les réunions de (d-2) droites passant par Q; ce point est alors d'ordre (d-1) sur C, ce qui contredit l'hypothèse.

Appendice B. Variétés de droites sauteuses d'un fibré stable de rang deux sur P^r.

Soit Σ un fibré stable de rang deux sur P^r. Notons E l'image réciproque de Σ sur la variété d'incidence F dans $G \times P^r$ (où G est la grassmannienne des droites de P^r).

Définition B.1 (Barth): <u>La variété des droites</u> i-<u>sauteuses pour</u> ε <u>est définie dans</u> G <u>par le</u> (i-1)-ième <u>idéal de Fitting de</u> $R^1q_*E(0,-[(c_1+2)/2])$ (<u>on utilise les mêmes notations qu'au paragraphe</u> 1, <u>de plus</u> c_1 <u>désigne ici la première classe de Chern de</u> Σ <u>et</u> $[\cdot]$ <u>la partie entière</u>).

Proposition B.2: <u>Le</u> <u>n-ième idéal de Fitting de</u> $R^1q_*E(0,m)$ <u>est égal au</u> (n-1)-<u>ième idéal de Fitting de</u> $R^1q_*E(0,m+1)$, <u>pour</u> $n,m \in Z$.

On peut évidemment supposer $m = 0$. Soit R l'anneau local de G en une droite. On sait que pour $k >> 0$, il existe une section de $\Sigma(k)$ n'ayant pas de zéro sur cette droite. Notant toujours E la restriction de E à $F \times_G \operatorname{spec} R = P^1_R$, celle-ci induit une suite exacte

$0 \longrightarrow \theta_{P^1_R}(-k) \longrightarrow E \longrightarrow 0_{P^1_R}(k') \longrightarrow 0$, avec $k'-k = c_1$. On en déduit pour $\ell \geq -k'-1$ une présentation

$$q_* \, \theta_{P^1_R}(\ell+k') \xrightarrow{\phi_\ell} R^1q_* \, \theta_{P^1_R}(\ell-k) \longrightarrow R^1q_*E(\ell) \longrightarrow 0 \, .$$

Choisissons une base (X_0,X_1) de $q_* \, \theta_{P^1_R}(1)$; pour tout s munissons $q_* \, \theta_{P^1_R}(s)$ de la base formée par les monômes de degré s en (X_0,X_1), et $R^1q_* \, \theta_{P^1_R}(-s-2)$ de la base duale. Soit (a_i) $(0 \leq i \leq k+k'-2)$ la matrice colonne, dans ces bases, de l'application $q_* \, \theta_{P^1_R} \longrightarrow R^1q_* \, \theta_{P^1_R}(-k-k')$. On montre que la matrice de ϕ_ℓ dans les bases choisies est $(a_{i+j})_{\substack{0 \leq i \leq k-\ell-2 \\ 0 \leq j \leq k'+\ell}}$. Pour cela, on considère le diagramme commutatif suivant

$$\begin{array}{ccc}
q_* \, \theta_{P^1_R} & \longrightarrow & R^1q_* \, \theta_{P^1_R}(-k-k') \\
\big\downarrow & & \big\downarrow \\
q_* \, \theta_{P^1_R}(\ell+k') & \longrightarrow & R^1q_* \, \theta_{P^1_R}(\ell-k)
\end{array}$$

où les deux flèches verticales sont déduites par fonctorialité de la multiplication par un monôme $X_0^i X_1^{\ell+k'-i}$, et on explicite les matrices de ces flèches dans les bases choisies. L'assertion est maintenant conséquence du lemme (2.3).

Proposition B.3: (I) <u>Si</u> $c_1(\Sigma)$ <u>est pair, la variété des droites</u> i-<u>sauteuses pour</u> Σ <u>est de codimension</u> $\leq 2i-1$ <u>lorsqu'elle n'est pas vide. Si sa codimension est</u> $2i-1$ <u>elle représente la classe de Chern</u>

$-c_{2i-1}\,(q_!\ E(0,i-1-(c_1+2)/2))$.

(II) <u>Si</u> $c_i(\Sigma)$ <u>est impair, la variété des droites</u> i-<u>sauteuses pour</u> Σ <u>est de codimension</u> $\leq 2i$ <u>lorsqu'elle n'est pas vide. Si sa codimension est</u> $2i$, <u>elle représente la classe de Chern</u>

$c_{2i}\,(q_!\ E(0,i-1-(c_1+1)/2))$.

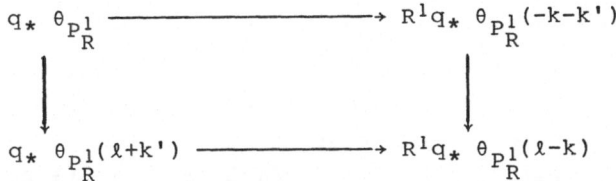

D'après B.2 , la variété des droites i-sauteuses est définie par le 0-ième idéal de Fitting de $R^1q_*(E(0,i-1-[(c_1+2)/2]))$. Considérons alors une présentation $0 \to N' \to N \to \varepsilon \to 0$ par des fibrés sur P^3 tels que $H^o(N \otimes_{\theta_{P^3}} \theta_L(i-1-[c_1+2)/2])) = 0$ pour toute droite L. Elle induit une suite exacte

$$0 \to q_*E(0,i-1-[(c_1+2/2]) \to P \to Q \to R^1q_*E(0,i-1-[(c_1+2)/2]) \to \to 0 ,$$

où P et Q sont des fibrés sur G. On a

$$rgP-rgQ = rg(q_*E(0,i-1-[(c_1+2/2])) = \begin{cases} 2i-2 & \text{si } c_1 \text{ est pair} \\ 2i-1 & \text{si } c_1 \text{ est impair} \end{cases}$$

ce qui démontre la proposition, compte tenu de (2.1 II).

Remarque B.4: Lorsque la codimension est maximale, la variété des droites i-sauteuses est donc "déterminantielle" et en particulier localement de Cohen-Macaulay.

Remarque B.5. Appliquant, comme dans le deuxième paragraphe le théoréme de Riemann-Roch-Grothendieck au morphisme $F \to G$, on peut par exemple démontrer:

1) Si Σ est un fibré sur P^3, avec $c_1(\Sigma) = 0$ et $c_2 = c_2(\Sigma)$, et si la variété des droites bisauteuses de Σ est une courbe, le degré de cette courbe est, comme Barth le suggère dans ([1]), $d = (c_2^3-c_2)/3$; de plus son genre arithmétique est $1+(c_2^3-c_2)(3c_2-8)/12 = 1+d(3c_2-8)/4$ lorsque Σ n'a qu'un nombre fini de droites trisauteuses.

2) Si Σ est un fibré sur P^4, avec $c_1(\Sigma) = 0$ et $c_2 = c_2(\Sigma)$, et si la variété des droites trisauteuses est une courbe, celle-ci est de degré $d = c_2(c_2+1)(c_2+4)(c_2^2-3c_2+4)/24$; son genre arithmétique est $1 + d(5c_2-15)/6$ s'il n'y a qu'un nombre fini de droites quadrisauteuses pour Σ.

3) Si Σ est un fibré sur P^4 tel que $c_1(\Sigma) = -1$, la variété des droites trisauteuses pour Σ est de degré $c_2(c_2+2)(c_2+6)[(c_2-2)^3+7(c_2-3)]/144$, lorsqu'elle est finie.

Note: Dans la démonstration de A.2, la récurrence repose aussi sur l'énoncé qui suit :

Proposition: Une adjointe d'ordre (i-1) à une première adjointe de C est une adjointe d'ordre i de C.

Ce résultat est une conséquence directe de l'assertion suivante qui peut se déduire de (1.5).

Lemme: Soient M et N deux modules de Cohen-Macaulay, de dimension d-1, de type fini, sur un anneau local régulier de dimension d. Si N est quotient d'une somme de copies de M, et si $F^0(M) = F^0(N)$, alors $F^i(N) \subset F^i(M)$ pour $i \geqslant 0$.

Bibliographie

[1] W. Barth. Counting singularities of quadratic forms on vector bundles. In Vector Bundles and Differential Equations. Progress in Matheamtics no 7, Birkhäuser.

[2] G. Kempf
D. Laksov. The determinental formula of Schubert calculus. Acta mathematica, vol. 132, 1973.

[3] P. Le Barz. Validité de certaines formules de géométrie énumerative. C.R. Acad. Sc. Paris, t. 289 (10 décembre 1979).

[4] R.L.E. Schwarzenberger. The secant bundle of a projective variety. Proc. of the London Math. Soc. Vol XIV, 1964.

[5] L. Szpiro. Travaux de Kempf, Kleiman, Laksov sur les diviseurs exceptionnels. Seminaire Bourbaki. Juin 1972, exp. no 417.

[6] L. Szpiro. Equations defining space curves. Tata Institute. Lectures on mathematics no 62 1979

Laurent GRUSON
Université de Lille
U.E.R. de Mathématiques
59655 - VILLENEUVE D'ASCQ
FRANCE

Christian PESKINE
Institute of Mathematics
P.O. Box 1053,
Blindern, OSLO 3
NORVEGE

SECTION PLANE D'UNE COURBE GAUCHE: POSTULATION

par

Laurent GRUSON et Christian PESKINE

Dans cette note, on voit qu'une simple transcription de la démonstration, fournie par Barth ([1]), du théorème de stabilité de Grauert-Mulich démontre le résultat suivant:

Lemme (Laudal): Soit C une courbe intègre de degré d de P^3 (1), non contenue dans une surface de degré s.

Si la section plane générale de C est contenue dans une courbe plane de degré s, on a $s^2+1 \geqslant d$.

Soient F la variété d'incidence dans $P^{3\vee} \times P^3$, et g et p ses projections sur $P^{3\vee}$ et P^3. Pour tout O_{P^3}-module M, nous noterons $M(m,n)$ le O_F-module $g^*(O_{P^{3\vee}}(m)) \otimes_{O_F} p^*(M(n))$. Si J est le faisceau d'idéaux de C dans P^3, nous avons par hypothèse $H^o(J(s)) = 0$ et $H^o(J(\alpha,s)) \neq 0$ pour $\alpha >> 0$. Nous pouvons évidemment supposer que s est minimum vérifiant cette propriété; soit alors α minimum tel qu'il existe une section non nulle de $J(\alpha,s)$. Cette section définit une hypersurface S de F, de bidegré (α,s), nécessairement réduite irréductible (sinon une de ses composantes réduites contiendrait l'image réciproque T de C dans F, ce qui contredirait

(1) Espace projectif de dimension 3 sur un corps algébriquement clos de caractéristique nulle.

l'hypothèse de minimalité faite sur s et α). Le morphisme $S \to P^3$ est alors dominant, génériquement lisse, et induit un complexe cotangent relatif exact:

$$0 \to O_S(-\alpha,-s) \to \Omega_{F/P^3} \otimes_{O_F} O_S \to \Omega_{S/P^3} \to 0.$$

Dualisant par rapport à O_S , nous obstenons une suite exacte

$$0 \to \Omega^{\vee}_{S/P^3} \to \Omega^{\vee}_{F/P^3} \otimes_{O_F} O_S \to O_S(\alpha,s).$$

Mais la dernière application de cette suite a son image contenue dans $J_S(\alpha,s)$ (ou J_S est l'idéal de T dans S) car le morphisme $S \to P^3$ n'est plat en aucun point de T . Remarquons de plus que l'application $\Omega^{\vee}_{F/P^3} \otimes_{O_F} O_S \to J_S(\alpha,s)$ a son conoyau de codimension au moins 2 dans F. Considérons maintenant sa restriction à une fibre générale de g, correspondant à un plan général H. Notons Γ la courbe de degré s restriction de S au plan H, et J_Γ l'idéal de la section plane $H \cap C$ dans Γ. Rappelons que le fibré projectif F sur P^3 est défini par $\Omega^{\vee}(-1)$, où Ω est le faisceau des différentielles sur P^3; on déduit facilement de la suite exacte $0 \to \Omega_{F/P^3} \to \Omega^{\vee}(-1,-1) \to O_F \to 0$ que la restriction de Ω_{F/P^3} à $g^{-1}(H)$ est $\Omega^{\vee}_H(-1)$. On obtient finalement une application $\Omega_H(1) \otimes_{O_H} O_\Gamma \to J_\Gamma(s)$ dont le conoyau est de codimension au moins 2 dans H. Soit alors I_Γ l'idéal de O_Γ, contenu dans J_Γ, tel que l'image de cette application soit $I_\Gamma(s)$, et soit N le fibré de rang 2 sur H défini par la suite exacte $0 \to N \to \Omega_H(1) \to I_\Gamma(s) \to 0$. Si Δ est le groupe de points de Γ dont l'idéal est I_Γ, et si $\delta = d^{\circ}\Delta$, un simple calcul sur les classes de Chern nous donne $s^2-\delta = c_2(N(s))-c_2(\Omega_H(1))$, soit $\delta = s^2+1-c_2(N(s))$. Mais $c_1(N) = -s-1$ entraine $c_2(N(s)) = c_2(N(1))$; comme il est clair que N n'a pas de section, l'énoncé est une conséquence immédiate du lemme suivant, appliqué a $N(1)$:

Lemme: Si M est un fibré de rang 2 sur P^2 tel que
$H^o(M(-1)) = 0$, alors $c_2(M) \geq 0$.

Si M est stable ou semi-stable, on sait que $c_2(M(t)) \geq 0$ pour tout t. Si M est instable, soit r le plus petit entier (positif) tel que $H^o(M(r)) \neq 0$; alors $c_1(M(r)) < 0$ (car M est instable) et $c_2(M(r)) \geq 0$, donc
$c_2(M) = c_2(M(r)) - rc_1(M(r)) + r^2 \geq 0$.

Remarque: Soit E un fibré de corrélation nulle sur P^3; si C est la courbe des zéros d'une section de $E(s)$, alors $d^oC = s^2+1$ et C n'est pas contenue dans une surface de degré s, mais toute section plane de C est sur une courbe de degré s (car toute section plane de E est un fibré semi-stable). On peut montrer que les courbes ainsi décrites sont les seules pour lesquelles l'inégalité de l'énoncé est une égalité.

Bibliographie

[1] W. Barth. Some properties of stable rank-2 vector bundles on P^n.
Math. Ann. 226, 125-150 (1977).

[2] O.A. Laudal. A generalized trisecant lemma.
Algebraic Geometry.
Lecture Notes in math. no. 687. Springer.

DEGENERATIONS OF COMPLETE TWISTED CUBICS

RAGNI PIENE

1. Introduction

Let $C \subset \mathbb{P}^3$ be a twisted cubic curve. Denote by $\Gamma \subset \mathrm{Grass}(1,3)$ its tangent curve (curve of tangent lines) and by $C^* \subset \check{\mathbb{P}}^3$ its dual curve (curve of osculating planes). The curve Γ is rational normal, of degree 4, while C^* is again a twisted cubic. The triple (C,Γ,C^*) is called a (non degenerate) <u>complete twisted cubic</u>. By a degeneration of it we mean a triple $(\bar{C},\bar{\Gamma},\bar{C}^*)$, where $\bar{C}$, $\bar{\Gamma}$, $\bar{C}^*$ are simultaneous flat specializations of C, Γ, C^*. Thus we work with Hilbert schemes rather than Chow schemes: let H denote the irreducible component of $\mathrm{Hilb}^{3m+1}(\mathbb{P}^3)$ containing the twisted cubics, $\check{H}$ the corresponding component of $\mathrm{Hilb}^{3m+1}(\check{\mathbb{P}}^3)$, and G the component of $\mathrm{Hilb}^{4m+1}(\mathrm{Grass}(1,3))$ containing the tangent curves of twisted cubics. The <u>space of complete twisted cubics</u> is the closure $T \subset H \times G \times \check{H}$ of the set of non degenerate complete twisted cubics.

In this paper we show how to obtain Schubert's 11 first order degenerations ([S],pp.164-166) of complete twisted cubics, viewed as elements of $H \times G \times \check{H}$, "via projections", i.e., by constructing 1-dimensional families of curves on various kinds of cones. In particular, we describe the ideals of the degenerated curves. A similar study was done by Alguneid [A], who viewed the degenerations as <u>cycles</u> (rather than flat specializations), and who gave equations for the complexes of lines associated to the degenerated cycles by using the theory of complete collineations.

An ultimate goal in the study of degenerations of complete twisted cubics, is of course to verify Schubert's results in the enumerative theory of twisted cubics. As long as one, as Schubert does, restricts oneself to only impose conditions that involve points, tangents, and osculating planes (and not secants, chords, osculating lines, ...), the space T is a compactification of the space of twisted cubics that contains enough information. In other words, one would like to describe the Chow ring of T in terms of cycles corresponding to degenerate complete twisted cubics, and in terms of cycles representing the various Schubert conditions. One approach would be to study the Chow ring of H and the blow-up map $T \to H$. In a joint work (to appear) with Michael Schlessinger we prove that the 12-dimensional scheme H is in fact smooth, and, moreover, that H intersects the other component H' (H' has dimension 15) of $\mathrm{Hilb}^{3m+1}(\mathbb{P}^3)$ transversally, along an 11-dimensional locus. (Note that H' contains plane cubic curves union a point in $\mathbb{P}^3$, and $H \cap H'$ consists of plane singular cubics with an embedded point "sticking out of" the plane.)

2. Degenerations via projections

Since all twisted cubics are projectively equivalent, we shall fix one, $C \subset \mathbb{P}^3 = \mathbb{P}^3_k$ (k algebraically closed field of characteristic 0), given by the ideal

$$I = (X_0X_2-X_1^2, X_1X_3-X_2^2, X_0X_3-X_1X_2).$$

Hence C has a parameter form

$$X_0 = u^3,\ X_1 = u^2v,\ X_2 = uv^2,\ X_3 = v^3.$$

The tangent curve Γ of C, viewed as a curve in $\mathbb{P}^5$ via the Plücker embedding of Grass(1,3), has a parameter form ($t = \frac{v}{u}$) given by the 2-minors of

$$\begin{pmatrix} 1 & t & t^2 & t^3 \\ 0 & 1 & 2t & 3t^2 \end{pmatrix},$$

hence by

$$Y_0 = v^4,\ Y_1 = 2uv^3,\ Y_2 = u^2v^2,$$
$$Y_3 = 3u^2v^2,\ Y_4 = 2u^3v,\ Y_5 = u^4.$$

The ideal of Γ in $\mathbb{P}^5$ is

$$J = (Y_3-3Y_2, 4Y_0Y_2-Y_1^2, Y_0Y_4-Y_1Y_2, Y_1Y_4-4Y_2^2, 4Y_0Y_5-Y_1Y_4, Y_1Y_5-Y_2Y_4, 4Y_2Y_5-Y_4^2).$$

The dual curve $C^* \subset \check{\mathbb{P}}^3$ has a parameter form given by the 3-minors of

$$\begin{pmatrix} 1 & t & t^2 & t^3 \\ 0 & 1 & 2t & 3t^2 \\ 0 & 0 & 1 & 3t \end{pmatrix}$$

hence by

$$\check{X}_0 = v^3, \ \check{X}_1 = 3uv^2, \ \check{X}_2 = 3u^2v, \ \check{X}_3 = u^3.$$

Since Γ is also equal to the tangent curve of C^* (under the canonical isomorphism Grass(lines in $\mathbb{P}^3$) $\cong$ Grass(lines in $\check{\mathbb{P}}^3$)) and C is the dual curve of C^* (see e.g. [P],§5), any type of degeneration $(\bar{C},\bar{\Gamma},\bar{C}^*)$ gives another type - called the dual degeneration - by reading the triple backwards.

Let $A \subset \mathbb{P}^3$ be a linear space, and choose a complement $B \subset \mathbb{P}^3$ of A. By projecting C onto B from the vertex A we obtain a degeneration of C: we construct a family $\{C_a\}$ of twisted cubics, contained in the cone of the above projection, over Spec $k[a]-\{0\}$. This family has a unique extension to a flat family over Spec $k[a]$, and the "limit curve" C_0 is thus a flat specialization of $C = C_1$ (see also [H], p.259, for the case A = a point). Note that interchanging the roles of A and B gives a limit curve equal to the curve C_∞ obtained in the similar way by letting $a \to \infty$, and C_∞ has the dual degeneration type of C_0. The type of degeneration obtained depends of course on the dimension and position of A and B w.r.t. C.

To find generators for the ideals of the degenerated curves, for chosen A and B, we start by writing down a parameter form of C_a, $a \neq 0$. (It is often convenient to introduce new coordinates at this point.) Then we determine enough generators for the ideal I_a of C_a, so that they specialize $(a = 0)$ to generators for the ideal I_0 of C_0. (Whenever C_0 acquires an embedded point, it turns out that a cubic generator is needed in addition to the (three standard) quadratic ones.)

The parameter form of C_a, $a \neq 0$, gives a parameter form of Γ_a, its tangent curve. As above we find generators for the ideal J_a of Γ_a that specialize to generators for J_0.

Similarly, one could work out the ideal of C_0^*. However, by a duality argument it is clear that C_0^* will have the degeneration type obtained (from C) by interchanging the roles of A and B. That is, C_0^* will be of the same type as C_∞, or, the degeneration type of C^* is equal to the dual degeneration type of C. For example, consider the degeneration type λ: A is a (general) point, B a (general) plane. When C degenerates along the cone over it, with vertex A, onto the plane B, its osculating planes degenerate towards the plane B. In $\check{\mathbb{P}}^3$, this means that C_a^* degenerates on the cone with vertex the <u>plane</u> $\check{A} \subset \check{\mathbb{P}}^3$ towards the <u>point</u> $\check{B} \in \check{\mathbb{P}}^3$. This degeneration type we call λ'; in general, we shall denote the dual degeneration by a "prime" in this way.

3. Schubert's 11 degenerations

We now give a list of Schubert's 11 types of degenerations, in his order and using his names for them.

λ A = general point (not on C, not on any tangent)

B = general plane (not osculating, not containing any tangent)

Take $A = (0,1,0,1)$, $B: X_3+X_1 = 0$,

and new coordinates:

$$X_0' = X_2+X_0,\ X_1' = X_3+X_1,\ X_2' = X_2-X_0,\ X_3' = X_3-X_1.$$

Then C_a, $a \neq 0$, is given by

$$X_0' = uv^2+u^3,\ X_1' = av^3+au^2v,\ X_2' = uv^2-u^3,\ X_3' = v^3-u^2v.$$

$$I_a = (a^2(X_0'-X_2')(X_0'+X_2') - (X_1'-aX_3')^2,$$
$$X_1'^2-a^2X_3'^2-a^2(X_0'+X_2')^2, -X_1'X_2'+aX_0'X_3',$$
$$(X_0'-X_2')(X_1'+aX_3')^2-a^2(X_0'+X_2')^3)$$

$$I_0 = (X_1'^2, X_1'X_3', X_1'X_2', X_3'^2(X_0'-X_2') - X_2'^2(X_0'+X_2'))$$
$$= ((X_3+X_1)^2, (X_3+X_1)(X_3-X_1), (X_3+X_1)(X_2-X_0),$$
$$X_0(X_3-X_1)^2 - X_2(X_2-X_0)^2)$$

Hence: C_0 is a plane nodal cubic with a nonplanar embedded point at the node.

λ' A = general plane
B = general point

Take A: $X_3+X_1 = 0$, $B = (0,1,0,1)$ and coordinates as for λ.

$$I_a = ((X_0'-X_2')(X_0'+X_2')-(aX_1'-X_3')^2,\ a^2X_1'^2-X_3'^2-(X_0'+X_2')^2,$$
$$X_0'X_3' - aX_1'X_2')$$

$$I_0 = (X_0'^2-X_2'^2-X_3'^2, X_3'^2+(X_0'+X_2')^2, X_0'X_3')$$
$$= (4X_0X_2-(X_3-X_1)^2,\ X_2(X_2+X_0), (X_3-X_1)(X_2+X_0))$$

Hence: C_0 is the union of three skew lines through the point $(0,1,0,1)$.

To find the degenerated tangent curve Γ_0 of λ (or of λ'):

Γ_a is given (in coordinates $Y_0',\ldots,Y_5'$ on $\mathbb{P}^5$ corresponding to $X_0',\ldots,X_3'$ on $\mathbb{P}^3$) on parameter form

$$Y_0' = v^4-2u^2v^2+u^4,\ Y_1' = 4auv^3,\ Y_2' = -av^4+4au^2v^2+au^4,$$
$$Y_3' = v^4+4u^2v^2-u^4,\ Y_4' = 4u^3v,\ Y_5' = av^4+2au^2v^2+au^4.$$
$$J_0 = (Y_2'-2Y_5', Y_1'^2, Y_1'Y_5', Y_5'^2, 12Y_3'Y_5'-7Y_1'Y_4', 7Y_0'Y_5'+11Y_3'Y_5',$$
$$Y_0'Y_1'-2Y_4'Y_5'-Y_1'Y_3', (4(Y_0'-Y_3')^2+12Y_3'(Y_0'-Y_3')-3Y_4'^2)^2$$
$$-(5Y_0'+4Y_3')^2((Y_0'-Y_3')^2+3Y_4'^2)).$$

Hence: Γ_0 is a plane tricuspidal quartic, with embedded points at the cusps.

κ A = point on a tangent, not on C
B = plane containing a tangent, not osculating.

Take $A = (0,1,0,0)$, B: $X_1 = 0$. Then C_a, $a \neq 0$, is given by

$$X_0 = u^3,\ X_1 = au^2v,\ X_2 = uv^2,\ X_3 = v^3,$$

$I_a = (a^2X_0X_2-X_1^2, X_1X_3-aX_2^2, aX_0X_3-X_1X_2, X_0X_3^2-X_2^3).$

$I_0 = (X_1^2, X_1X_2, X_1X_3, X_0X_3^2-X_2^3).$

Hence: C_0 is a cuspidal cubic, in the plane $X_1 = 0$, with a nonplanar embedded point at the cusp.

κ' A = plane containing a tangent, not osculating.

B = point on tangent, not on C.

$I_a = (X_0X_2-a^2X_1^2, aX_1X_3-X_2^2, X_0X_3-aX_1X_2).$

$I_0 = (X_0X_2, X_2^2, X_0X_3).$

Hence: C_0 is the union of the line $X_2 = X_3 = 0$ with the double line $X_0 = X_2 = 0$ (doubled on a quadratic cone with vertex $(0,1,0,0)$).

The tangent curve Γ_a of κ (or κ') is given by

$Y_0 = v^4,\ Y_1 = 2auv^3,\ Y_2 = au^2v^2,$

$Y_3 = 3u^2v^2,\ Y_4 = 2u^3v,\ Y_5 = av^4.$

$J_0 = (Y_2, Y_1^2, Y_1Y_3, Y_1Y_4, Y_1Y_5, Y_0Y_5, Y_2Y_5, 27Y_0Y_4^2-4Y_3^3).$

Hence: Γ_0 is a cuspidal cubic, in the plane $Y_1 = Y_2 = Y_5 = 0$, with a nonplanar embedded point at the cusp $(1,0,0,0,0,0)$ (this point corresponds to the flex tangent of C_0), union the line $Y_0 = Y_1 = Y_2 = Y_3 = 0$, intersecting the cubic in its flex $(0,0,0,0,1,0)$ (corresponding to the cusp tangent of C_0).

ω A = point on C

B = osculating plane

Take $A = (0,0,0,1)$, B: $X_3 = 0$.

C_a, $a \neq 0$, is given by

$X_0 = u^3,\ X_1 = u^2v,\ X_2 = uv^2,\ X_3 = av^3,$

$I_a = (X_0X_2-X_1^2, X_1X_3-aX_2^2, X_0X_3-aX_1X_2).$

$I_0 = (X_0X_2-X_1^2, X_0X_3, X_1X_3).$

Hence: C_0 is the union of a conic, in the plane $X_3 = 0$, with the line $X_0 = X_1 = 0$.

ω' A = osculating plane

B = point on C

$I_a = (X_0X_2-X_1^2, aX_1X_3-X_2^2, aX_0X_3-X_1X_2).$

$I_0 = (X_0X_2-X_1^2, X_1X_2, X_2^2).$

Hence: C_0 is the triple line $X_1 = X_2 = 0$ (tripled on a quadratic cone with vertex $(0,0,0,1)$).

The tangent curve Γ_a of ω (or ω') is given by

$Y_0 = av^4,\ Y_1 = 2auv^3,\ Y_2 = u^2v^2,\ Y_3 = 3au^2v^2,$

$Y_4 = 2u^3v,\ Y_5 = u^4.$

$J_0 = (Y_3, Y_1^2, Y_1Y_4, Y_1Y_5, Y_0Y_5, Y_0Y_4-Y_1Y_2, 4Y_2Y_5-Y_4^2).$

Hence: Γ_0 is the union of a conic, in the plane $Y_0 = Y_1 = Y_3 = 0$, with the double line $Y_1 = Y_3 = Y_4 = Y_5 = 0$.

θ To obtain this degeneration, we choose A to be a "line-plane" (L,U), s.t. for some $x \in C$, $x \in L \subset U$, $tg_x \subset U$, $L \neq tg_x$, U not osculating - and B a "point-line" (P,L'), s.t. for some $x \in C$, $L' \subset osc_x$, $\{P\} = L' \cap tg_x$, $x \neq P$, $P \notin C$. Then we form a 2-dimensional family $\{C_{a,b}\}$, where the parameter a corresponds to projecting C from U to P, and b to projecting from L to L'. Taking a = b we obtain a 1-dimensional family $\{C_{a,a}\}$.

Take $L: X_1-X_3 = X_2 = 0$, $U: X_2 = 0$ and $L': X_0 = X_1+X_3 = 0$, $P = (0,0,1,0)$. In new coordinates $X_0, X_1' = X_1-X_3$, $X_2, X_3' = X_1+X_3$, $C_{a,b}$ is given by $X_0 = abu^3$, $X_1' = au^2v-av^3, X_2 = uv^2$, $X_3' = abu^2v+abv^3$

$$I_{a,b} = (4abX_0X_2-(X_3'+bX_1')^2, (X_3'+bX_1')(X_3'-bX_1')-4a^2b^2X_2^2,$$
$$X_0(X_3'-bX_1')-ab(X_3'+bX_1')X_2, X_0(X_3'-bX_1')^2-4a^3b^3X_2^3).$$

By letting a = b, rewriting the generators, and letting a = 0, we obtain

$$I_0 = (X_0(X_1+X_3), (X_1-X_3)(X_1+X_3), (X_1+X_3)^2, X_0((X_1-X_3)^2-X_0X_2)).$$

<u>Hence</u>: C_0 is the union of a conic, in the plane $X_1+X_3=0$, with its tangent line at $(0,0,1,0)$, and with that point as a nonplanar embedded point.

θ' A = "point-line"

B = "line-plane".

$$I_{a,b} = (4bX_0X_2-a(bX_3'+X_1')^2, a^2(bX_3'+X_1')(bX_3'-X_1') -4X_2^2, abX_0(bX_3'-X_1')-(bX_3'+X_1')X_2).$$

Taking $a = b$ and $a = 0$ gives

$$I_0 = (4X_0X_2-(X_1-X_3)^2, (X_1-X_3)X_2, X_2^2)$$

<u>Hence</u>: C_0 is the line $X_1-X_3 = X_2 = 0$ tripled on a quadratic cone with vertex $(1,0,1,0)$.

The tangent curve Γ_a of θ (or θ') (for $a = b$) is given by

$$Y_0' = av^4-au^2v^2,\ Y_1' = 4a^2uv^3,\ Y_2' = v^4+u^2v^2,$$

$$Y_3' = 3a^3u^2v^2+a^3u^4,\ Y_4' = 2au^3v,\ Y_5' = -3a^2u^2v^2+a^2u^4.$$

$$J_0 = (Y_3', Y_0'Y_1', Y_1'^2, Y_1'Y_4', Y_1'Y_5', Y_0'Y_5', 4Y_2'Y_5'-Y_0'^2-Y_4'^2).$$

<u>Hence</u>: Γ_0 is the union of a conic, in the plane $Y_0' = Y_1' = Y_3' = 0$, with the two lines $Y_1' = Y_3' = Y_5' = Y_0'^2+Y_4'^2 = 0$, with the common point of intersection, $(0,0,1,0,0,0)$, as an embedded point (this point corresponds to the line - tangent to the conic - of C_0).

δ A = a line, not contained in any osculating plane, and intersecting C in exactly one point.

B = a line, not intersecting C, contained in exactly one osculating plane.

Take A: $X_0-X_2 = X_1 = 0$, B: $X_0+X_2 = X_3 = 0$ and change coordinates: $X_0' = X_0-X_2, X_1, X_2' = X_0+X_2, X_3$. Then C_a is given by

$$X_0' = u^3-uv^2,\ X_1 = u^2v,\ X_2' = au^3+auv^2,\ X_3 = av^3.$$

$$I_a = (X_2'^2-a^2X_0'^2-4a^2X_1^2, 4aX_1X_3-(X_2'-aX_0')^2, (X_2'+aX_0')X_3-aX_1(X_2'-aX_0')).$$

By changing the generators, we see

$$I_0 = ((X_0+X_2)^2,(X_0+X_2)X_3,X_0^2-X_2^2-2X_1X_3).$$

<u>Hence</u>: C_0 is the union of the line $X_0+X_2 = X_1 = 0$ with the double line $X_0+X_2 = X_3 = 0$, and is contained in a smooth quadric.

δ' $I_a = (a^2X_2'^2-X_0'^2-4X_1^2, 4aX_1X_3-(aX_2'-X_0')^2,$
$a(aX_2'+X_0')X_3-X_1(aX_2'-X_0'))$

$$I_0 = ((X_0-X_2)^2,(X_0-X_2)X_1,X_1^2) = (X_0-X_2,X_1)^2$$

<u>Hence</u>: C_0 is the triple line $X_0-X_2 = X_1 = 0$ (tripled by taking its 1st order neighbourhood in $\mathbb{P}^3$).

The tangent curve Γ_a for δ (or δ') is given by

$Y_0' = a^2v^4+3a^2u^2v^2$, $Y_1' = 2auv^3$, $Y_2' = au^2v^2-au^4$,
$Y_3' = -av^4+3au^2v^2$, $Y_4' = 4au^3v$, $Y_5' = u^2v^2+u^4$.

$J_0 = (Y_0', Y_1'(2Y_1'+Y_4'), Y_2'(2Y_1'+Y_4'), Y_4'(2Y_1'+Y_4'),$
$Y_2'^2+Y_1'^2, Y_4'(Y_3'-2Y_2'), Y_2'(Y_3'-2Y_2'))$

<u>Hence</u>: Γ_0 is the union of the two lines
$Y_0' = 2Y_1'+Y_4' = Y_3'-2Y_2' = Y_1'^2+Y_2'^2 = 0$, with the double line $Y_0' = Y_1' = Y_2' = Y_4' = 0$.

η A = general line, i.e. $A \cap C = \emptyset$,
A not contained in an osculating plane

B = general line (same conditions as for A, since these are self-dual!)

Take A: $X_0-X_3 = X_1+X_2 = 0$, B: $X_0+X_3 = X_1-X_2 = 0$, and change coordinates:

$X_0' = X_0-X_3$, $X_1' = X_1-X_2$, $X_2' = X_1+X_2$, $X_3' = X_0+X_3$. Then C_a is given by

$X_0' = u^3-v^3$, $X_1' = au^2v-auv^2$, $X_2' = u^2v+uv^2$,
$X_3' = au^3+av^3$.

$$I_a = ((aX_0'+X_3')(aX_2'-X_1')-(X_1'+aX_2')^2,(X_1'+aX_2')(X_3'-aX_0')$$
$$-(aX_2'-X_1')^2,X_3'^2-a^2X_0'^2-a^2X_2'^2+X_1'^2)$$
$$I_0 = (X_1'^2,X_1'X_3',X_3'^2) = (X_1-X_2,X_0+X_3)^2$$

Hence: C_0 is the tripled line $X_1-X_2 = X_0+X_3 = 0$ (tripled as in δ').

η' is of the same type as η, since the conditions on A, B are self dual.

The tangent curve Γ_a of η is given by

$$\begin{aligned}
Y_0' &= av^4+2auv^3-2au^3v-au^4\\
Y_1' &= -a^2v^4+2a^2uv^3+2a^2u^3v-a^2u^4\\
Y_2' &= 2au^2v^2\\
Y_3' &= 6au^2v^2\\
Y_4' &= v^4+2uv^3+2u^3v+u^4\\
Y_5' &= -av^4+2auv^3-2au^3v+au^4 .
\end{aligned}$$

$$J_0 = (Y_3'-3Y_2',Y_0'Y_1',Y_1'^2,Y_1'Y_2',Y_1'Y_5',Y_0'Y_5'-Y_1'Y_4'+3Y_2'^2,$$
$$4Y_1'Y_4'-6Y_0'Y_5'+3Y_0'^2-Y_5'^2)$$

Hence: Γ_0 is the union of four lines in the three-space $Y_3'-3Y_2' = Y_1' = 0$, with an embedded point (sticking out of that space) at their common point of intersection.

Remark: By choosing other A's and B's we can obtain further types of degenerations. For example, consider the degeneration obtained by taking A = a chord of C , B = an axis of C (i.e., the intersection of two osculating planes). Then C_0 is the union of three skew lines, meeting in 2 points, whereas its dual is a triple line (1st order nbhd. of a line in $\check{\mathbb{P}}^3$). The tangent curve Γ_0 is the union of two double lines.

On the next page, we give a figure showing Schubert's 11 degenerate complete twisted cubics. Each triple should also be read backwards!

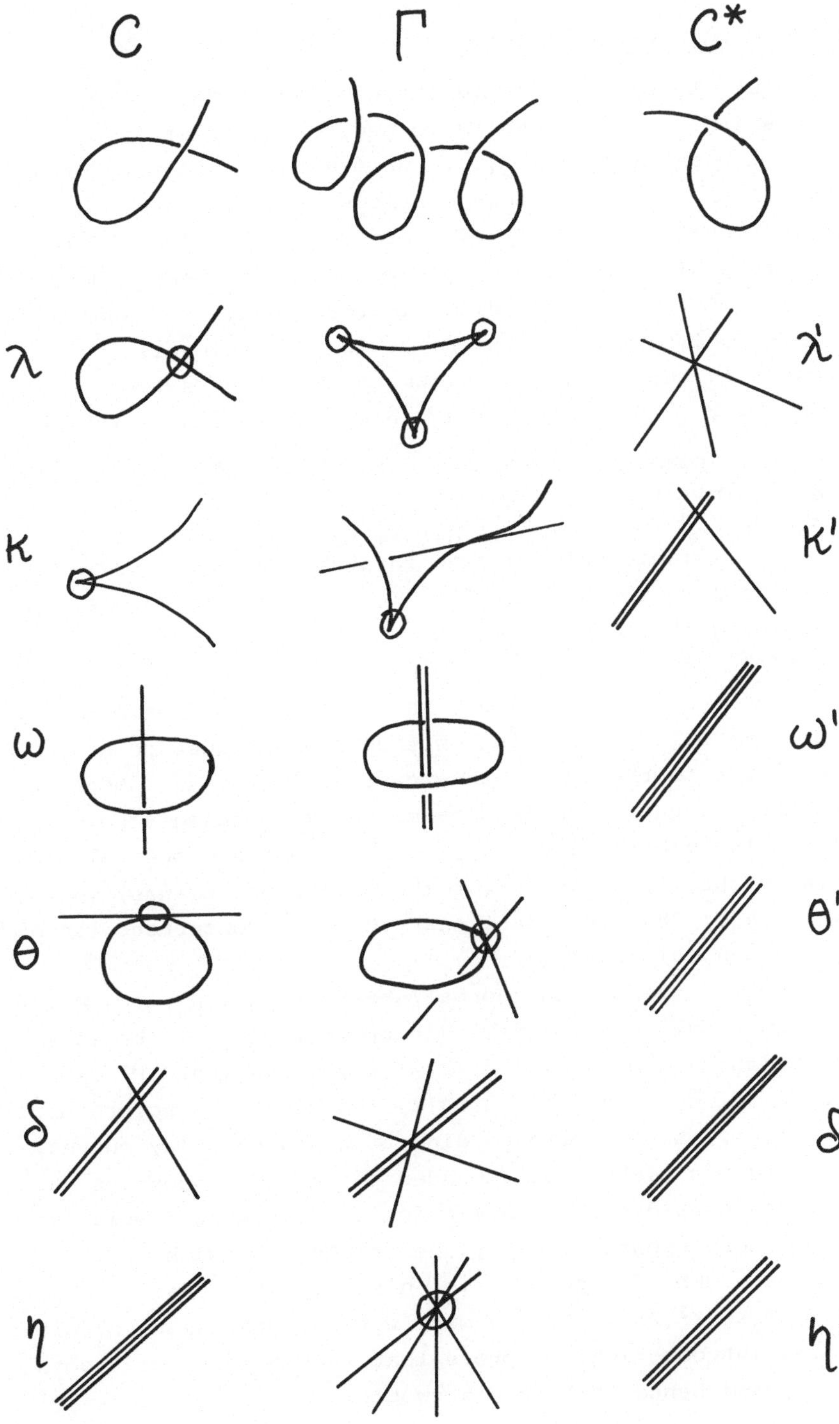
C
Γ
C*
λ
λ'
κ
κ'
ω
ω'
θ
θ'
δ
δ'
η
η'

4. Some remarks on H and T

Let $T_\lambda, T_\kappa, \ldots$ denote the closure of the set of points in T corresponding to degenerations of type $\lambda, \kappa, \ldots$, and let $H_\lambda, H_\kappa, \ldots$ denote the similarly defined sets in H. That the degenerations $\lambda, \kappa, \ldots$ are of first order, means that $T_\lambda, T_\kappa, \ldots$ are of codimension 1 in T; this is easily seen to be true by counting the parameters of each of the corresponding figures. Only H_λ and H_ω are of codimension 1 in H, so the (birational) projection map $\pi: T \to H$ blows up the other sets $H_\kappa, H_{\lambda'}, \ldots$. For example, H_κ has codimension 2 (there are ∞^{10} plane cuspidal cubics in $\mathbb{P}^3$), and for a given $\bar{C} \in H_\kappa$, $\pi^{-1}(\bar{C}) =$

$$\left\{(\bar{C},\bar{\Gamma},\bar{C}^*);\ \bar{\Gamma} = \text{a (uniquely determined) cuspidal cubic union a line through the flex}\right\}$$

Since "a line through the flex" corresponds to "a plane containing the cusp tangent of $\bar{C}$", we see that $\dim \pi^{-1}(\bar{C}) = 1$.

The set H_η ($=H_{\delta'}$) has the largest codimension, namely 8; all degenerations without an embedded point specialize to these. In this case, $\pi^{-1}(\bar{C})$ has dimension 7: the tangent curve is determined by choosing 4 point-planes through the line $\bar{C}_{red}$, which satisfy one relation between the cross-ratios (of the points and planes) (see e.g. [A],p.206, or recall that the four concurrent lines $\bar{\Gamma}_{red}$ span only a $\mathbb{P}^3$).

Let N denote the normal sheaf of $\bar{C} \in H_\eta$ in $\mathbb{P}^3$. One can prove, e.g. by taking a presentation of the ideal of $\bar{C}$, that $\dim H^0(N,\bar{C}) = 12$. It follows that H is smooth at $\bar{C}$, since $\dim H = 12$, and hence all points of $H - H_\lambda$ (i.e., those corresponding to Cohen-Macaulay curves, i.e., curves without an embedded point) are smooth on H.

Now consider H_λ. Any point in it can be specialized to one corresponding to a plane triple line with a nonplanar embedded point, e.g. given by the ideal $(X_1X_3, X_2X_3, X_3^2, X_1^3)$. In the work with M. Schlessinger, cited in the introduction, we prove that such a point is smooth on H, and hence that H is smooth.

<u>Remark</u>: The results $\dim H^0(N,\bar{C}) = 12$ if $\bar{C} \in H_\eta$, and $\dim H^0(N,\bar{C}) = 16$ if $\bar{C}$ is a plane triple line with embedded point, have also been obtained by Joe Harris; he also gives a list of possible degeneration types of a curve $C \in H$ (private communication).

As a final comment, let us mention an advantage of working with Hilbert schemes rather than Chow schemes: the existence of universal families of curves, which allows the following way of expressing Schubert's various conditions as cycles on T. Namely, let

$$\begin{array}{ccc} \mathcal{C} \subset \mathbb{P}^3\times T & \quad \mathcal{C}' \subset \mathrm{Grass}(1,3)\times T & \quad \mathcal{C}^* \subset \check{\mathbb{P}}^3\times T \\ p \downarrow \;\swarrow & \quad q\downarrow \;\swarrow & \quad r\downarrow\;\swarrow \\ T & \quad T & \quad T \end{array}$$

denote the universal families (pulled back to T from H, G, $\check{H}$ respectively). The condition, denoted ν by Schubert, for a curve C to intersect a given line L, is then represented by the cycle $T_\nu = p_*(\mathcal{C} \cap L\times T)$; the condition, Schubert's ρ, that the curve touches a given plane U, by $T_\rho = q_*(\mathcal{C}' \cap \sigma_{1,1}\times T)$, where $\sigma_{1,1}$ is the 2-plane in Grass(1,3) of lines in U, and so on. We plan to return to the question of determining the relations between these cycles and the cycles $T_\lambda, T_\kappa, \ldots$ - and to a study of the Chow ring of T.

Acknowledgments. This work was begun while the author was a visitor of the Department of Mathematics, Universidade Federal de Pernambuco, in March/April 1981. Many thanks are due to Prof. I. Vainsencher for conversations that started the work, and to CNPq for its financial support.

<u>Bibliography</u>

[A] A.R. Alguneid, "Analytical degeneration of complete twisted cubics", Proc. Cambridge Phil.Soc. 52(1956), 202-208.

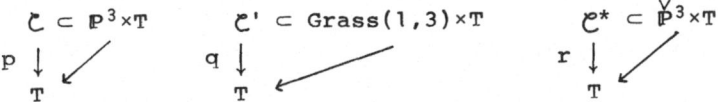

[H] R. Hartshorne, <u>Algebraic Geometry</u>. New York-Heidelberg -Berlin, Springer-Verlag 1977.

[P] R. Piene, "Numerical characters of a curve in projective n-space". In <u>Real and complex singularities</u>, <u>Oslo 1976</u>. Ed. P. Holm. Groningen: Sijthoff and Noordhoff 1978; pp. 475-495.

[S] H. Schubert, <u>Kalkül der abzählenden Geometrie</u>. B.G. Teubner, Leipzig 1879. (New edition: Springer-Verlag, 1979.)

Matematisk institutt, P.B. 1053 Blindern, Oslo 3, Norway.

PLURICANONICAL - GORENSTEIN - CURVES

Fabrizio Catanese

§ 0. Introduction

Some of the most classical results in algebraic geometry deal with the pluricanonical mappings of a complete smooth curve.

Classically, if X is a smooth curve of genus $p \geq 2$ the sections of $H^o(X, (\Omega_X^1)^{\otimes n})$ give the nth pluricanonical map Φ_n : $X \to \mathbb{P}(H^o(X, (\Omega_X^1)^{\otimes n})^v)$ of X into the projective space associated to the dual of $H^o(X, (\Omega_X^1)^{\otimes n})$.

It is well-known that this map is an embedding if

i) $n \geq 3$

ii) $n = 2$ and $p \geq 3$

iii) $n = 1$, $p \geq 3$ and X is not hyperelliptic.

Indeed, iii) characterizes hyperelliptic curves because if X is hyperelliptic then the canonical map Φ_1 yields a double cover of a rational normal curve. The need for extending these results to singular and reducible curves appears if one studies families of smooth curves and the possible degenerations of the generic fibre.

The first result in this direction, namely the extension of i) above for certain curves with nodes, was proved by Deligne and Mumford ([4]) in their work on the irreducibility of the moduli space of curves of genus p. F. Sakai ([10]) encountered similar problems in his study of open surfaces and his work shows the usefulness of having results of this kind for reduced curves lying on a smooth surface.

The object of this paper is to investigate in more detail this problem, with a greater generality which we hope suffices for most applications.

The correct analogue of pluricanonical mappings in the case of reducible curves with singularities is obtained replacing the sheaf Ω^1_X of 1-forms by the dualizing sheaf ω_X ; in order that sections of this sheaf, or tensor powers of it, define a map to some projective space, one has to assume that ω_X be an invertible sheaf (cf. [8], lecture 5). Notice that ω_X is invertible if and only if X is Gorenstein. We call the mapping associated to the linear system $|\omega_X^{\otimes n}|$ the nth pluricanonical mapping of X.

In order for these mappings to be well defined on X one should not have components of X along which all sections of $\omega_X^{\otimes n}$ vanish. This leads naturally to the following definition.

Definition 0.1: X is said to be semi-canonically positive (S.C.P. for short) if and only if, for each component Y of X, the degree of $\omega_X|_Y = \omega_X \otimes \mathcal{O}_Y$ is non negative. If, for each Y, this degree is positive, X is said to be canonically positive (C.P.).
It is clear that if some pluricanonical mapping is an embedding, then X must be C.P.

Now we can state our simplest results.

Theorem A: If X is S.C.P $|\omega_X^{\otimes n}|$ is base point free for each $n \geq 2$.

Theorem B: If X is C.P. $|\omega_X^{\otimes n}|$ gives an embedding of X for each $n \geq 3$.

We shall also study in more detail the structure of the maps associated with ω_X and $\omega_X^{\otimes 2}$ and indeed the greater part of this paper will be devoted to prove the analogue of ii), and of iii) above under suitable conditions of connectedness. We shall also show that our conditions of connectedness are close to being necessary and sufficient for the validity of our statements and we shall produce several explicit examples.

This paper is organized as follows: In § 1 we recall known basic facts about Gorenstein curves, we show how to obtain a S.C.P. Gorenstein curve out of an arbitrary Gorenstein curve by destroying some components that we call negative tails, and we describe S.C.P. curves of genus one. In § 2 we discuss the behavior of $|\omega_X^{\otimes n}|$, for $n \geq 2$, using Riemann Roch duality and some explicit interpretations of first cohomology groups: we prove the above Theorems A and B, and also (Theorem C) describe when $|\omega_X^{\otimes 2}|$ does not give an embedding. § 3 is devoted to the study of the canonical map $|\omega_X|$ and, in particular, we describe explicitly the "hyperelliptic curves" (the ones for which the canonical map is not birational). Finally, in § 4 we show by means of an example that even the simplest Theorems A and B do not carry over to the non-reduced case without additional hypotheses.

Our notation is as follows:

k is an algebraically closed field over which all the varieties in question are defined.

If V is a k-vector space, V^{v} is its dual.

If X is a projective scheme, with structure sheaf $\mathcal{O}_X$, ω_X is the dualizing sheaf of X (see [7] p. 242); moreover, if $\mathcal{F}$ is a coherent sheaf on X, we denote by $\mathcal{F}^* = \mathcal{H}om_{\mathcal{O}_X}(\mathcal{F}, \mathcal{O}_X)$, by $h^i(\mathcal{F})$ the dimension of $H^i(X, \mathcal{F})$ as a k-vector space, by

$$\chi(\mathcal{F}) = \Sigma(-1)^i h^i(\mathcal{F}).$$

If X is a reduced curve, $X = Y \cup Z$, with $\dim(Y \cap Z) = 0$, we will denote Z by $X - Y$. Also, $Y \cdot Z$ is defined to be equal to the length of $\mathcal{O}_{Y \cap Z}$, and if $x \in X$, $(Y \cdot Z)_x$ is, by definition, the length of $\mathcal{O}_{Y \cap Z, x}$.

If Y is a subscheme of X, $\mathcal{F}$ is coherent on X, $\mathcal{F}|_Y$ stands for $\mathcal{F} \otimes \mathcal{O}_Y$.

If s is a section of $\mathcal{F}$, $s \equiv 0$ means that the stalk of s is 0 at any point of X; $\equiv$ is also used to denote linear equivalence of divisors. Without explicit mention we shall assume all the schemes under consideration to be complete.

R. R. is an abbreviation for the Grothendieck-Serre-Riemann-Roch duality theorem (see [7], [11]) which, in the case of curves, reads out as follows:

$$\mathrm{Hom}\ (\mathcal{F}, \omega_X)^{v} \cong H^1\ (X, \mathcal{F}),\quad \mathrm{Ext}^1\ (\mathcal{F}, \omega_X)^{v} \cong H^0\ (X, \mathcal{F}).$$

The arithmetic genus $p\ (X)$ of a curve X is, by definition, equal to $1 - \chi\ (\mathcal{O}_X)$, and then $\deg\ (\omega_X) = 2 - p\ (X)$. If X is Gorenstein $\Phi_n : X \to \mathbb{P}(H^0\ (X, \omega_X^{\otimes n})^{v})$ is the nth - (pluri) canonical map of X.

§ 1. Gorenstein Curves

Lemma 1.1: Let W be a projective variety (possibly non connected), F an invertible sheaf on W, $\mathcal{G}$ a torsion free sheaf on W such that for each invertible sheaf L on W $\mathrm{Hom}\ (L, F) \cong \mathrm{Hom}\ (L, \mathcal{G})$: then $F \cong \mathcal{G}$.

Proof: Assume W to be connected. Then $\mathrm{Hom}\ (F, F) = k$ hence there is a non trivial homomorphism $a : F \to \mathcal{G}$. a must be injective since $\mathcal{G}$ is torsion free, hence, if $K = \mathrm{coker}\ a$, we have, upon tensoring with $\mathcal{O}_W\ (n)$, the following exact sequence

$$0 \to F\ (n) \to \mathcal{G}(n) \to K\ (n) \to 0$$

For n large enough $H^i\ (W, F\ (n)) = 0$ for $i \geq 1$, and

$H^0(\mathcal{G}(n)) \cong H^0(F(n))$ by our hypothesis, since e.g. $H^0(F(n)) = \mathrm{Hom}(\mathcal{O}_W(-n), F)$. Then $H^0(K(n)) = 0$ for all n large enough, hence $K = 0$ and $F \cong \mathcal{G}$. If W is not connected, it suffices to show that if Y is a connected component of W, then our hypothesis holds true for Y, $F|_Y$, $\mathcal{G}|_Y$. But, for any invertible sheaf L' on Y, consider L on W to be equal to L' on Y, and to $\mathcal{O}_W(n)$ on $W-Y$: then for n large enough $\mathrm{Hom}(L', F|_Y) \cong \mathrm{Hom}(L,F)$, and the same is true for $\mathcal{G}$.

Q.E.D.

<u>Proposition 1.2</u> (<u>Noether's Formula</u>): Let $\pi : Y \to X$ be a finite birational morphism of Gorenstein curves. Then $\omega_Y \cong \pi^*(\omega_X) \otimes \tilde{C}$, where $\tilde{C}$ is the conductor of π viewed as an ideal sheaf on Y.

<u>Proof</u>: By the previous lemma, it suffices to prove that, for every invertible sheaf L on Y,

$$\mathrm{Hom}(L, \omega_Y) \cong \mathrm{Hom}(L, \pi^*(\omega_X) \otimes \tilde{C}).$$

Taking the dual vector spaces, the left hand side is $H^1(Y,L)$, while the right hand side is

$$H^0(Y, \mathcal{H}om_{\mathcal{O}_Y}(L, \pi^*(\omega_X) \otimes \tilde{C}))^{\vee}.$$

The map being finite, $H^1(Y,L) = H^1(X, \pi_*L) = \mathrm{Hom}(\pi_*L, \omega_X)^{\vee}$, therefore it is enough to show that

$$\pi_*(\mathcal{H}om_{\mathcal{O}_Y}(L, \pi^*(\omega_X) \otimes \tilde{C})) \cong \mathcal{H}om_{\mathcal{O}_X}(\pi_*L, \omega_X).$$

This equality indeed is of local nature and follows from the fact that, at the finite set of points x where π is not an isomorphism, the conductor ideal $C = \pi_*\tilde{C}$ is equal to $\mathcal{H}om_{\mathcal{O}_X}(\pi_*\mathcal{O}_Y, \mathcal{O}_X)$, by its very definition.

Q.E.D.

Assume now that X is Gorenstein and reduced, and that $Y = \tilde{X}$ is the normalization of X.

We have then the standard exact sequence

$$(1.3) \qquad 0 \to \mathcal{O}_X \to \pi_* \mathcal{O}_{\tilde{X}} \to \Delta \to 0$$

where $\Delta = \bigoplus_{x \text{ singular}} \Delta_x$ and one usually denotes by δ_x the length of Δ_x, by δ the one of Δ $(\delta = \Sigma_x \delta_x)$.

Applying the functor $\mathcal{H}om_{\mathcal{O}_X}(\ ,\mathcal{O}_X)$ one obtains the dual exact sequence

$$(1.3') \qquad 0 \to C \to \mathcal{O}_X \to \mathcal{E}xt^1_{\mathcal{O}_X}(\Delta,\mathcal{O}_X) \to$$

$$\to \mathcal{E}xt^1_{\mathcal{O}_X}(\pi_*\mathcal{O}_{\tilde{X}},\mathcal{O}_X) \to 0$$

Let $M = \mathcal{O}_{X/C}$, $m_x = \text{length } (M_x)$.

<u>Lemma 1.4</u>: $C^* = \pi_* \mathcal{O}_{\tilde{X}}$, hence $\Delta \cong \mathcal{E}xt^1(M, \mathcal{O}_X)$.

<u>Proof</u>: For any ideal $\mathcal{J} \subset \mathcal{O}_X$ of finite colength, $\mathcal{J}$ contains a non zero divisor f. Then $\mathcal{J}^* \subset K_X$, where K_X is the full ring of fractions of $\mathcal{O}_X$: in fact if $\psi \in \mathcal{J}^*$, $h \in \mathcal{J}$, $\psi(h) f = \psi(hf) = h\psi(f)$, and thus $\psi(h) = h \ (\psi(f)/_f)$ and we can write $\psi = (\psi(f)/_f) \in K_X$.

Consider now $\psi \in C^*$: since C is an ideal in $\pi_*\mathcal{O}_{\tilde{X}}$, $\forall g \in \pi_*\mathcal{O}_{\tilde{X}}$, $\forall f \in C$, we have $\psi g f \in \mathcal{O}_X$, then $\psi f \in C$.

So $\psi C \subset C$, and ψ is regular on $\tilde{X}$. The last statement follows by taking the dual of the exact sequence

$$0 \to C \to \mathcal{O}_X \to M \to 0$$

Q.E.D.

<u>Theorem 1.5</u>: ω_X is invertible at x if and only if the following equivalent conditions are satisfied:

a) $\mathcal{E}xt^1(\pi_* \mathcal{O}_{\tilde{X}}, \mathcal{O}_X) = 0$

b) $\delta_x = m_x$

c) for each coherent sheaf F with supp $(F) = x$,

length $(\mathcal{E}xt^1(F, \mathcal{O}_X))$ = length (F).

Morever, in general, if x is singular $1 \le m_x \le \delta_x$.

Proof: We defer the reader to Serre's book ([11]pp. 76-80) for the proof of the more difficult parts, a, b $\Rightarrow \omega_X$ invertible, $m_x \le \delta_x$. We shall prove instead that ω_X invertible $\Rightarrow$ c $\Rightarrow$ a,b.

In fact, if ω_X is invertible at x, then $\mathcal{E}xt^1(F, \mathcal{O}_X) \cong \mathcal{E}xt^1(F, \omega_X)$, hence its length is $h^0(\mathcal{E}xt^1(F, \omega_X)) = \dim(\mathrm{Ext}^1(F, \omega_X)) = h^0(F)$ by R.R.

If c) holds, by virtue of the exact sequence

$$0 \to M \to \mathcal{E}xt^1(\Delta, \mathcal{O}_X) \to \mathcal{E}xt^1(\pi_* \mathcal{O}_{\tilde{X}}, \mathcal{O}_X) \to 0$$

and of lemma 1.4., one obtains

$$\delta_x = m_x \quad \text{and} \quad \mathcal{E}xt^1(\pi_* \mathcal{O}_{\tilde{X}}, \mathcal{O}_X) = 0$$

Q.E.D.

Let $\tilde{X}_1, \ldots \tilde{X}_k$ be the connected (irreducible) components of $\tilde{X}$; then the long exact cohomology sequence associated to (1.3) gives

$$(1.6) \qquad (k-1) \le \delta \quad , \quad p(X) = \Sigma_{h=1}^{k} p(\tilde{X}_h) + (\delta-k+1).$$

Actually, if X' is the disjoint union of the irreducible components X_i of X, one has an exact sequence analogous to (1.3)

$$(1.7) \qquad 0 \to \mathcal{O}_X \to \pi'_* \underset{\mathcal{O}_{X'}}{\overset{k}{\underset{i=1}{\oplus}} \mathcal{O}_{X_i}} \to \Delta' \to 0$$

One associates to X a graph $|X|$ in the following way: take a segment $|X_i|$ for each component X_i, and mark a point $|y_i|$ in $|X_i|$ for each singular point y of X belonging to X_i; then, if

X_i and X_j meet at y, identify $|y_i|$ with $|y_j|$.

<u>Proposition 1.8</u>: Let X be a reduced connected curve; then $p(X) = 0$ if and only if

a) every component X_i is isomorphic to $\mathbb{P}^1$,

b) the singularities of X are given by r smooth branches with independent tangents,

c) the associated graph $|X|$ is contractible.

Moreover, if $p(X) = 0$, X is Gorenstein iff it has only nodes as singularities.

<u>Proof</u>: By (1.7) $p(X) = 0 \Longrightarrow p(X_i) = 0$. If X is irreducible, by (1.6), $p(X) = 0$ implies $p(\tilde{X}) = 0$, $\delta = 0$, hence X is smooth and $\cong \mathbb{P}^1$. One can assume clearly that X_1 is such that $Y = X - X_1$ is connected. Let Z be the disjoint union of Y and X_1, and consider the obvious morphism $p : Z \to X$. Again one has the exact sequence

$$0 \to \mathcal{O}_X \to p_* \mathcal{O}_Z \to \Delta'' \to 0,$$

therefore Δ'' has length 1, hence, first of all, Y and X_1 intersect in a single point y. Then $\mathcal{O}_{X,y}$ is a subring of $\mathcal{O}_{X_1,y} \oplus \mathcal{O}_{Y,y}$ contained in the subring $R = \{(f,g) \mid f(y) = g(y)\}$: since however Δ'' has length 1, $R = \mathcal{O}_{X,y}$, hence

$$\dim \mathfrak{m}_{y,Y}/\mathfrak{m}^2_{y,Y} + \dim \mathfrak{m}_{y,X_1}/\mathfrak{m}^2_{y,X_1} = \dim \mathfrak{m}_{y,X}/\mathfrak{m}^2_{y,X}$$

and b),c) are proven by induction on k.

The converse is also easy.

If X is Gorenstein at x, $\delta_x + 1$ components meet transversally at x, but here $C = \mathfrak{m}_x$, so $\delta_x = m_x = 1$, and x is thus a node.

Q.E.D.

<u>Definition 1.9</u>: X is said to be m - connected if, for each decomposition $X = Y \cup Z$, with $\dim Y \cap Z = 0$, one has $Y \cdot Z \geq m$. Y and Z are said to meet transversally at $x \in X$ if $(Y \cdot Z)_x = \dim \mathcal{O}_{Y \cap X, x} = 1$.

Recall now that we are assuming X to be connected, hence always 1-connected : if X is not 2-connected, then one can write $X = Y \cup Z$ with Y and Z intersecting transversally at a single point x.

<u>Proposition 1.10</u>: If $X = Y \cup Z$ and Y and Z meet transversally at x, then x is a node for X if ω_X is invertible at x.

<u>Proof</u>: The question is local, but, taking a normalization of X at the other points of intersection of Y and Z, and at the points where ω_X is not invertible, we can assume that X be Gorenstein and that $Y \cap Z = \{x\}$. Let $\pi : \tilde{X} \to X$ be the normalization of X at x, $\tilde{Z} = \pi^{-1}(Z)$, $\tilde{Y} = \pi^{-1}(Y)$.

We have the following commutative diagram

$$\begin{array}{ccccccccc} 0 & \to & \mathcal{O}_X & \to & \mathcal{O}_Z & \oplus & \mathcal{O}_Y \to & \mathcal{O}_{Z \cap Y} & \to 0 \\ & & \downarrow & & & \downarrow & & & \\ 0 & \to & \pi_* \mathcal{O}_{\tilde{X}} & \to & \pi_* \mathcal{O}_{\tilde{Z}} & \oplus & \pi_* \mathcal{O}_{\tilde{Y}} \to 0 & & \end{array}$$

Therefore, if $\Delta_X = \pi_* \mathcal{O}_{\tilde{X}}/\mathcal{O}_X$, $\delta_X = h^0(\Delta_X)$, and δ_Z, δ_Y are defined in the same way, $\delta_X = 1 + \delta_Z + \delta_Y$. Let C_X, C_Y, C_Z be the conductor ideals of the morphisms π, $\pi|_{\tilde{Y}}$, $\pi|_{\tilde{Z}}$, and let m_X, m_Y, m_Z be defined accordingly.

If Y, Z are smooth at x our claim is proven, otherwise, since $\mathcal{m}_{x,X} = \mathcal{m}_{x,Y} + \mathcal{m}_{x,Z}$, and $C_X \subset \mathcal{m}_{x,X}$, if they are both singular $C_X = C_Y \oplus C_Z$, if Y is smooth $C_X = C_Z \oplus \mathcal{m}_{x,Y}$; in the first case $m_X = m_Y + m_Z - 1$, in the second $m_X = m_Z$.

In either case we have a contradiction, since $\delta_X = m_X, m_Z \leq \delta_Z$, $m_Y \leq \delta_Y$.

Q.E.D.

We are now going to show how a Gorenstein curve can fail to be S.C.P. (cf 0.1).

<u>Proposition 1.11</u>: Let $Y \subset X$ be a connected union of components X_i of X such that $\deg \omega_X|_{X_i} \leq 0$. Then $p(Y) = 0$ unless $Y = X$, $\deg \omega_X|_{X_i} = 0$ for each component X_i. In this last case $p(X) = 1$ and $\omega_X \cong \mathcal{O}_X$. Conversely, if X is S.C.P. and $p(X) = 1$, ω_X is trivial.

<u>Proof</u>: Y being connected, $p(Y) = h^1(Y, \mathcal{O}_Y) =$ (by R.R.) $=$ dim $\mathrm{Hom}_{\mathcal{O}_X}(\mathcal{O}_Y, \omega_X) = h^0(X, \mathcal{J}_{X-Y}\,\omega_X)$ where $\mathcal{J}_{X-Y}$ is the ideal sheaf of $X-Y$: in fact, by the exact sequence

$$0 \to \mathcal{J}_Y \to \mathcal{O}_X \to \mathcal{O}_Y \to 0$$

$$0 \to \mathcal{H}om_{\mathcal{O}_X}(\mathcal{O}_Y, \mathcal{O}_X) \to \mathcal{O}_X \to \mathcal{H}om_{\mathcal{O}_X}(\mathcal{J}_Y, \mathcal{O}_X)$$

thus $\mathcal{H}om_{\mathcal{O}_X}(\mathcal{O}_Y, \mathcal{O}_X) = \{f \in \mathcal{O}_X \mid f \cdot g = 0 \ \forall g \in \mathcal{J}_Y\} = \mathcal{J}_{X-Y}$.
Assume now that $X-Y \neq \emptyset$: then every section s of $\mathcal{J}_{X-Y}\ \omega_X$ is identically zero on $X-Y$, so vanishes at some point of Y; but then, by the assumption made on $\deg \omega_X|_{X_i}$ for $X_i \subset Y$, s is identically zero. The same clearly holds if $\exists\, X_i \subset Y$ s.t. $\deg \omega_X|_{X_i} < 0$. Assume then $X = Y$; $p(X) = 0$ unless $\deg \omega_X|_{X_i} = 0$ for each i, and conversely if $p(X) = 0$, by prop. 1.8, X is not S.C.P. But then $p(X) = h^0(\omega_X) \geq 1$, so there exists a non zero section $s : \mathcal{O}_X \to \omega_X$. Since $\deg \omega_X = 2\,p(X) - 2$, if $\deg \omega_X|_{X_i} = 0$ for each i, s gives an

isomorphism of ω_X with the trivial sheaf, and if $p(X) = 1$, X is S.C.P., then $\deg \omega_X|_{X_i} = 0$ for each i.

Q.E.D.

Lemma 1.12: Let $Z \subset X$ be both Gorenstein. Then the ideal $\mathcal{J}_{X-Z} \otimes \mathcal{O}_Z$ is invertible, $\omega_X|_Z = \omega_Z \otimes (\mathcal{J}_{X-Z} \otimes \mathcal{O}_Z)^{-1}$, and in particular $\deg \omega_{X|Z} = \deg \omega_Z + Z \cdot (X-Z)$

Proof: Let W be a normalization of $X-Z$ at the points where ω_{X-Z} is not invertible, let Y be the disjoint union of W and Z, $\pi : Y \to X$ the natural map. By proposition 1.2 we get that $\omega_Z = \pi^* \omega_X \otimes \tilde{C} = \omega_X \otimes C \otimes \mathcal{O}_Z$, and that $C \otimes \mathcal{O}_Z$ is invertible.

We have the following commutative diagram

$$
\begin{array}{ccccccccc}
 & & & & 0 & & & & \\
 & & & & \uparrow & & & & \\
 & & & & \Delta_{X-Z} & & & & \\
 & & & & \uparrow & & & & \\
0 & \to & \mathcal{O}_X & \to & \mathcal{O}_Z \oplus \pi_* \mathcal{O}_W & \to & \Delta_X & \to & 0 \\
 & & \uparrow \wr & & \uparrow & & \uparrow & & \\
0 & \to & \mathcal{O}_X & \to & \mathcal{O}_Z \oplus \mathcal{O}_{X-Z} & \to & \mathcal{O}_{Z \cap (X-Z)} & \to & 0 \\
 & & & & \uparrow & & \uparrow & & \\
 & & & & 0 & & 0 & &
\end{array}
$$

and, by dualizing (taking $\mathcal{H}om_{\mathcal{O}_X}(\ ,\mathcal{O}_X)$) we obtain

$$
\begin{array}{ccccc}
 & & 0 & & \\
 & & \downarrow & & \\
0 & \to & C = \mathcal{J}_{X-Z} \oplus C_W & \to & \mathcal{O}_X \\
 & & \downarrow & & \downarrow \wr \\
0 & \to & \mathcal{J}_{X-Z} \oplus \mathcal{J}_Z & \to & \mathcal{O}_X
\end{array} .
$$

Therefore $C \otimes \mathcal{O}_Z = \mathcal{J}_{(X-Z)} \otimes \mathcal{O}_Z$: finally we have an exact sequence

$$
0 \to \mathcal{J}_{X-Z} \otimes \mathcal{O}_Z \to \mathcal{O}_Z \to \mathcal{O}_{X-Z} \otimes \mathcal{O}_Z \to 0
$$
$$
\mathcal{O}_{X-Z} \otimes \mathcal{O}_Z = \mathcal{O}_{(X-Z) \cap Z}
$$

and the last assertion follows then.

Q.E.D.

<u>Proposition 1.13</u>: Let X be Gorenstein, $\pi : \tilde{X} \to X$ the normalization of a singular point x, Z an irreducible component of X containing x, $\tilde{Z} = \pi^{-1}(Z)$, $\delta_Z = h^o(\pi_* \mathcal{O}_{\tilde{Z}} / \mathcal{O}_Z)$. Then $\deg(\tilde{C}^{-1}|_{\tilde{Z}}) = 2\delta_Z + Z\,(X-Z)_x$.

<u>Proof</u>: We can clearly assume, as in 1.10, that $Z \cap (X-Z) = x$. We have then the following exact commutative diagram

$$\begin{array}{ccccccccc}
 & & 0 & & 0 & & & & \\
 & & \downarrow & & \downarrow & & & & \\
0 & \to & \mathcal{O}_X & \to & \mathcal{O}_Z \oplus \mathcal{O}_{X-Z} & \to & \mathcal{O}_{Z\cap(X-Z)} & \to & 0 \\
 & & \downarrow & & \downarrow & & & & \\
 & & \pi_* \mathcal{O}_{\tilde{X}} & \cong & \pi_* \mathcal{O}_{\tilde{Z}} \oplus \pi_* \mathcal{O}_{(\tilde{X}-\tilde{Z})} & & & & \\
 & & \downarrow & & \downarrow & & & & \\
0 \to \mathcal{O}_{Z\cap(X-Z)} & \to & \Delta_X & \to & \Delta_Z \oplus \Delta_{X-Z} & \to & 0 & & \\
 & & \downarrow & & \downarrow & & & & \\
 & & 0 & & 0 & & & &
\end{array}$$

Taking the dual sequences we obtain

$$\begin{array}{ccccccc}
 & 0 & & 0 & & & \\
 & \downarrow & & \downarrow & & & \\
 & \pi_*(\tilde{C}|_{\tilde{Z}}) \oplus \pi_*(\tilde{C}|_{\tilde{X}-\tilde{Z}}) & \cong & C & & & \\
 & \downarrow & & \downarrow & & & \\
0 \to & \mathcal{I}_{X-Z} \oplus \mathcal{I}_Z & \to & \mathcal{O}_X & \to & \mathcal{E}xt^1(\mathcal{O}_{Z\cap(X-Z)}, \mathcal{O}_X) & \to 0 \\
 & \downarrow & & \downarrow & \nearrow & & \\
0 \to & \mathcal{E}xt^1(\Delta_Z, \mathcal{O}_X) \oplus \mathcal{E}xt^1(\Delta_{X-Z}, \mathcal{O}_X) & \to & \mathcal{E}xt^1(\Delta_X, \mathcal{O}_X) & & & \\
 & \downarrow & & \downarrow & & & \\
 & 0 & & 0 & & &
\end{array}$$

therefore $\text{length}\,(\mathcal{O}_{\tilde{Z}} / \tilde{C}|_{\tilde{Z}}) = \text{length}\,(\pi_* \mathcal{O}_{\tilde{Z}} / \mathcal{O}_Z) + \text{length}\,(\mathcal{O}_Z / \mathcal{I}_{X-Z}) +$

$+ \text{ length } (\mathcal{I}_{X-Z/\pi_*(\tilde{C}|\tilde{Z})}) = \delta_Z + Z \ (X-Z)_x +$

$+ \text{ length } (\mathcal{E}xt^1 (\Delta_Z, \mathcal{O}_X)) = 2\ \delta_Z + Z\ (X-Z)_x$ by 1.5 c).

Q.E.D.

Remark 1.14: Passing to the completion of the local rings in consideration, and considering branches of Z through x, one obtains an entirely analogous result (with the same proof) for the multiplicity of $\tilde{C}$ at a point q of $\tilde{X}$ s.t. $\pi(q) = x$ (see also [6] for a slightly different proof).

Definition 1.15: A negative tail Y contained in X is a maximal connected curve in X s.t. $\deg \omega_{X|Y} < 0$, and s.t., for each curve $X_i \subset Y$, $\deg \omega_{X|X_i} \leq 0$.

Proposition 1.16: A negative tail Y is Gorenstein with $p(Y) = 0$, and Y intersects $X-Y$ transversally at a single point.

Proof: $p(Y) = 0$ by 1.11, and if Y is Gorenstein, by 1.12 $\deg \omega_{X|Y} = -2 + \dim \mathcal{O}_{Y \cap (X-Y)} < 0$, hence $Y \cdot (X-Y) = 1$. Since a smooth $\mathbb{P}^1$ is Gorenstein, and $\deg \omega_{X|X_i} < 0$ for some component $X_i \subset Y$, there exists a maximal connected $Y' \subset Y$ such that Y' is Gorenstein, $\deg \omega_{X|Y'} < 0$: we claim that $Y' = Y$. In fact Y' intersects $X-Y'$ transversally in a point x which belongs to a component W of X : if $W \not\subset Y$, then Y' is a connected component of Y, hence $Y' = Y$, otherwise W is a smooth $\mathbb{P}^1 \subset Y$, and $Y' \cup W$ is Gorenstein $\subset Y$, a contradiction.

Q.E.D.

Remark 1.17: Let X be, as usual, reduced and Gorenstein: then X is S.C.P iff X contains no negative tails. By throwing away the negative tails, one can obtain from any X a new connected Gorenstein curve $X' \subset X$ which is S.C.P. It is clear that any section of ω_X^n, for $n \geq 1$, vanishes identically on the negative tails.

Let $x_1, \dots x_k$ be the nodes where X' intersects $X-X'$: then by 1.12

$$\omega_X^{\ n}|_{X'} = \omega_{X'}^{\ n} \ (\Sigma_{i=1}^{k} n\ x_i) \quad , \text{ and } \quad H^o\ (X, \omega_X^{\ n}) \cong$$

$$\cong H^o\ (X', \omega_{X'}^{n}\ ((n-1)\ \Sigma_{i-1}^{\ k}\ x_i)) \quad .$$

Then the rational maps $|\omega_X|$ and $|\omega_{X'}|$ coincide on X', while, for $n \geq 2$, $|\omega_{X'}^{n}|$ is obtained by $|\omega_X^{\ n}|$ followed by a projection.

To end this section, let me describe the S.C.P. Gorenstein curves X with $p\ (X) = 1$.

<u>Proposition 1.18</u>: A S.C.P. Gorenstein curve X with $p\ (X) = 1$ belongs to the following classes:

a1) - a5) (X lies on a smooth surface)

a1) X smooth, a2) X is rational with a node

a3) X is rational with an ordinary cusp a4) X consists of 2 $\mathbb{P}^1$'s tangent at a point (X tacnodal)

a5) X has only nodes and is a cycle of $\mathbb{P}^1$'s

b) X consists of k smooth $\mathbb{P}^1$'s meeting in a point x where the tangents to the branches are linearly dependent, but any (k-1) of them are independent.

<u>Proof</u>: By (1.7) $\sum_{i=1}^{k} p\ (X_i) \leq 1$, if $X_1, \dots X_k$ are the irreducible components of X.

If $p\ (X_1) = 1$, since, by proposition 1.10, $\omega_X \cong \mathcal{O}_X$, if X_1 is Gorenstein , $\omega_{X_1} \cong \mathcal{O}_{X_1}$ and by 1.12 $X = X_1$. In fact X_1 is Gorenstein by

Lemma 1.19: An irreducible curve Y with $p(Y) = 1$ is Gorenstein and belongs to one of the classes a1), a2), a3).

Proof: Assume Y to be singular, and let $\pi : \tilde{Y} \to Y$ be the normalization. Then $\delta = 1$, hence Y has only one singular point x, and 1.5 implies that $1 \le m_x \le \delta_x = 1$, so Y is Gorenstein.
Then, since $\pi_* \mathcal{O}_{\tilde{Y}/C}$ has dimension 2, $\tilde{C}$ has degree -2; therefore either $\pi^{-1}(x) = p$, or $\pi^{-1}(x) = \{p_1, p_2\}$. In both cases $\mathcal{O}_{x,X} = k \oplus \mathfrak{m}_{x,X} = k \oplus C_{x,X}$ as a subring of $\pi_* \mathcal{O}_{\tilde{Y}}$, therefore in the first case x is an ordinary cusp, in the second x is a node.

Q.E.D.

End of proof of 1.18: Assume then that each X_i is a smooth $\mathbb{P}^1$. Then, by 1.12 $X_i(X-X_i) = 2$. Assume that X_1 intersects $X-X_1$ in 2 points (which are therefore nodes) : then it is easy to see that the same must hold for all X_i's (in fact there exists a maximal $Y \subset X$ such that X has only nodes along Y, and if $Y \neq X$, $\exists\, W \subset X$, $W \not\subset Y$, s.t. $W \cap Y \neq \emptyset$: then W intersects $X-Y-W$ at a point which is not a node, hence $W(X-W) \ge 3$, a contradiction). It is now obvious that the graph associated to X is a cycle, so that we are in case a5).

Otherwise we have that all the X_i's intersect in a single point x.

Then $\delta_x = m_x = k$: so $\mathfrak{m}_{x,X}$ has codimension 1 in $\mathfrak{m}' = \oplus_{i=1}^{k} \mathfrak{m}_{x,X_i}$ and, by Nakayama's lemma, $\mathfrak{m}^2_{x,X} = \oplus_{i=1}^{k} \mathfrak{m}^2_{x,X_i}$; in particular $C \supset \mathfrak{m}^2_{x,X}$, but then equality holds since $k = \dim \mathcal{O}_{x,X}/C = \dim \mathcal{O}_{x,X}/\mathfrak{m}^2_{x,X}$

Let t_i be a uniformizing parameter for X_i at x, e_i the function which is 1 on X_i and 0 on the other X_j's.
We know that $\mathfrak{m}_{x,X}/\mathfrak{m}^2_{x,X}$ is an hyperplane in $\oplus_{i=1}^{k} \mathfrak{m}_{x,X_i}/\mathfrak{m}^2_{x,X_i}$, so there exist $\alpha_i \in k$ s.t.

$$\mathfrak{m}_{x,X}/\mathfrak{m}^2_{x,X} = \left\{ \sum_{i-1}^{k} a_i t_i \;\middle|\; \Sigma a_i \alpha_i = 0 \right\} .$$

Clearly, also, $C/\mathfrak{m}^2_{x,X} = \left\{ \sum_{i=1}^{k} b_i t_i \;\middle|\; b_j \alpha_j = 0 \ \forall j \right\}$

since $f \in C$ iff $f \cdot e_j \in \mathcal{O}_{x,X}$ $\forall j = 1, \dots k$.

The conclusion is that every α_j is $\neq 0$.

Q.E.D.

Remark 1.20: If X is S.C.P. and has only nodes as singularities, one has (cf. [4]) zero tails, i.e. chains of $\mathbf{P}^1$'s over which ω_X is trivial, and $|\omega_X^n|$ contracts these zero tails to points, so that the image of X is just the image of a C.P. X' obtained by taking off these tails and setting together the 2 "end points" of the tail to build a node.

If Y is a smooth $\mathbf{P}^1$ tangent to X-Y at a smooth point x, one can throw away Y and obtain X' by putting a cusp in x (i.e. if t is a uniformizing parameter at x, one replaces $\mathcal{O}_{x,X-Y}$ by the subring generated by $1, t^2, t^3$). Analogously if Y crosses X-Y in a node, one replaces the node by a tacnode to get X'.

Thus, there is also a natural way to obtain from a S.C.P X a C.P. X', such that $|\omega_X^n|$ has the same image of $|\omega_{X'}^n|$.

We won't however use this construction.

§ 2. The Pluricanonical Maps

To prove the first results (theorems A, B, C) we have to show the vanishing of some first cohomology groups: in turn, using R.R. duality, these are interpreted as certain homomorphisms, over which a rough hold is given by the following lemmas.

Lemma 2.1: Let x be a singular point of X, $\tilde{X}$ the normalization of X at x, $\hat{X}$ the blow-up of the maximal ideal $\mathcal{M}_x$, $\pi: \tilde{X} \to X$, $p: \hat{X} \to X$ the natural maps. Then $\mathcal{H}om_{\mathcal{O}_X}(\mathcal{M}_x, \mathcal{O}_X)$ is naturally embedded in $\pi_* \mathcal{O}_{\tilde{X}}$, and actually in the subsheaf $p_* \mathcal{O}_{\hat{X}}$.

Proof: x being a singular point, $\mathcal{M}_x \supset C$, hence there is a natural map $\mathcal{H}om_{\mathcal{O}_X}(\mathcal{M}_x, \mathcal{O}_X) \to \mathcal{H}om_{\mathcal{O}_X}(C, \mathcal{O}_X) = \pi_* \mathcal{O}_{\tilde{X}}$.

The fact that this map is injective follows either from the arguments of lemma 1.4 or from the sharper statement that this sheaf embeds in $p_* \mathcal{O}_{\hat{X}}$. Let $f_1 \dots f_r$ be elements of $\mathcal{O}_{x,X}$ which induce a basis of $\mathcal{M}_{x,X}/\mathcal{M}^2_{x,X}$, and such that f_i is not a 0-divisor.

Let $\psi \in \mathcal{H}om_{\mathcal{O}_X}(\mathcal{M}_x, \mathcal{O}_X)$. Assume that $\psi(f_i) = \varphi_i$: then, $\forall f \in \mathcal{M}_{x,X}$ we have $\psi(f f_i) = f \psi(f_i) = f_i \psi(f)$, hence $\psi(f) = f \varphi_i/f_i$ (we are working in the full ring K_x of fractions of $\mathcal{O}_{x,X}$). The first thing to remark is that φ_i cannot be a unit, otherwise $f_j = \varphi_j f_i \varphi_i^{-1}$, contradicting the independence of the f_i's mod $\mathcal{M}^2_{x,X}$.

But then ψ is given by multiplication by the rational function $\varphi_i/f_i = \varphi_j/f_j$ which is easily seen to be regular on $\hat{X}$.

Q.E.D.

Lemma 2.2: $\pi: \tilde{X} \to X$ being as in the previous lemma, let M be the (invertible) sheaf of ideals generated by $\pi^{-1}(\mathcal{M}_x)$: then $\mathcal{H}om_{\mathcal{O}_X}(\mathcal{M}^2_x, \mathcal{O}_X)$ embeds in $\pi_*(M^{-1})$.

Proof: If $\mathcal{I}$ is an ideal which contains a non 0-divisor h, we have seen (1.4) that $\mathcal{H}om_{\mathcal{O}_X}(\mathcal{I}, \mathcal{O}_X)_x = \{ g \in K_x \mid g \cdot \mathcal{I} \subset \mathcal{O}_X \}$:

If $g \cdot \mathfrak{m}_x^2 \subset \mathcal{O}_X$, $\forall f' \in \mathfrak{m}_x$, $(g \cdot f')\ \mathfrak{m}_x \subset \mathcal{O}_X$, so, by 2.1., $g \cdot f' \subset \pi_* \mathcal{O}_{\tilde{X}}$, hence $g \in \pi_* (M^{-1})$.

Q.E.D.

Theorem A: Let X be a S.C.P. reduced Gorenstein curve, n an integer ≥ 2: then $|\omega_X{}^n|$ is free from base points.

Proof: By 1.11, we can clearly assume $p(X) \geq 2$. Consider the standard exact sequence (k_x being the residue field at x)

(2.3) $$0 \to \mathfrak{m}_x \omega_X{}^n \to \omega_X{}^n \to k_x \to 0.$$

By R.R. $H^1(\omega_X{}^n) = H^0(\omega_X{}^{1-n}) = 0$ since X is S.C.P and $p(X) \geq 2$. Then x is not a base point iff $H^1(\mathfrak{m}_x\ \omega_X{}^n) = 0$. By R.R., again, we have to show that

$$\mathrm{Hom}_{\mathcal{O}_X}(\mathfrak{m}_x, \omega_X{}^{1-n}) = 0$$

Assume now that x is a singular point of X. Let $\pi : \tilde{X} \to X$ be the normalization at x : by 2.1 it suffices to show that

$H^0(\tilde{X}, \pi^*(\omega_X{}^{1-n})) = 0$.

Let $\tilde{Y}$ be a connected component of $\tilde{X}$; on each irreducible component $\tilde{Y}_i$ of $\tilde{Y}$ $\pi^*(\omega_X) \otimes \mathcal{O}_{\tilde{Y}_i}$ has degree ≥ 0, so it is enough to prove that for some i this degree is > 0.

But if this were not to hold, $Y = \pi(\tilde{Y})$ would have $p(Y) = 0$ by 1.10 and $\omega_X|_Y$ would be trivial : hence Y would be contained in the base locus of $|\omega_X{}^n|$, in particular there would be smooth base points. Let's then prove that a smooth point x cannot be a base point. Denote by Z the irreducible component to which x belongs, and set for commodity $F = \omega_X^{1-n} \otimes \mathfrak{m}_x^{-1}$. Since $p(X) \geq 2$, $\deg F = 1 + 2(1-n)(p-1) < 0$, hence clearly $H^0(X,F) = 0$ if $\deg \omega_X|_Z \geq 1$. On the other hand $\deg \omega_X|_Z = 0 \Rightarrow p(Z) = 0$ (1.11). By 1.12 $\dim \mathcal{O}_{Z \cap (X-Z)} = 2$, and, since $Z \cong \mathbb{P}^1$,

$$H^0(Z, F|_Z) = H^0(Z, \mathcal{O}_Z(x)) \to H^0(\mathcal{O}_{Z \cap X-Z} \otimes F)$$

is an isomorphism.

By the exact cohomology sequence associated to the sequence

$$0 \to F \to F|_Z \oplus F|_{X-Z} \to F \otimes \mathcal{O}_{Z \cap X-Z} \to 0$$

we obtain $H^0(X, F) = H^0(X-Z, F|_{X-Z}) = H^0(X-Z, \omega_X^{1-n}|_{X-Z})$.

By the previous argument his vector space is 0 if on every connected component Y of X-Z $\deg \omega_X|_Y > 0$. But Z (X-Z) = 2 implies that there are at most 2 connected components.

If X-Z is connected, clearly $\deg \omega_X|_{X-Z} > 0$.

If X-Z has two connected components Y_1, Y_2 (thus meeting Z transversally at two distinct points Y_1, Y_2) and say, $\deg \omega_X|_{Y_1} = 0$, then $p(Y_1) = 0$, hence $\deg \omega_X|_{Y_1} = -1$, an evident contradiction.

Q.E.D.

<u>Definition 2.4</u>: An elliptic tail of a C.P. curve X is an irreducible component Y of X such that $p(Y) = 1$, Y (X-Y) = 1.

<u>Theorem B</u>: If X is C.P. $|\omega_X^n|$ gives an embedding of X for $n \geq 3$.

<u>Theorem C</u>: If $p(X) = 2$ or X has elliptic tails $|\omega_X^2|$ does not give a birational map Φ_2. If X is C.P. , $p(X) \neq 2$, X has no ellip-

tic tails, Φ_2 is an embedding unless (possibly, cf. 3.23) at a point x if

a) $X \supset W$ with $W \cap (X-W) = \{x\}$, $W \cdot (X-W) = 2$, $p(W) = 1$, W is either rational irreducible with a cusp at x or a cycle of $2\mathbf{P}^1$'s meeting at x, and morever $C \not\subset \mathfrak{m}^2_{x,X}$

b) $X \supset W \cong \mathbf{P}^1$, $W \cap (X-W) = \{x\}$, $W \cdot (X-W) = 3$.

<u>Proof of Theorems B,C</u>: Let x,y be 2 points of X and consider the exact sequence

$$(2.5) \qquad 0 \to \mathfrak{m}_x \mathfrak{m}_y \, \omega_X{}^n \to \omega_X{}^n \to \omega_X{}^n/\mathfrak{m}_x \mathfrak{m}_y \, \omega_X{}^n \to 0$$

where, if $x = y$, $\mathfrak{m}_x \mathfrak{m}_y$ has to be understood as $\mathfrak{m}_x{}^2$. Since, for $n \geq 2$, $H^1(X, \omega_X{}^n) = 0$, $|\omega_X{}^n|$ gives an embedding if and only if

$$(2.6) \qquad \mathrm{Hom}_{\mathcal{O}_X}(\mathfrak{m}_x \mathfrak{m}_y, \omega_X{}^{1-n}) = 0 \quad .$$

We have to consider separately the following cases:

i) x, y smooth

ii) x singular, y smooth

iii) $x \neq y$, x, y both singular

iv) $x = y$ singular

<u>i</u>: Let F be the invertible sheaf $\omega_X{}^{1-n} \otimes (\mathfrak{m}_x \mathfrak{m}_y)^{-1}$: we have to prove that $H^0(X,F) = 0$. For every component X_i of X, $\deg F|_{X_i} = (1-n) \deg \omega_X|_{X_i} + \rho$, where $\rho = 2$ if $x, y \in X_i$, $\rho = 0$ if $x, y \notin X_i$, $\rho = 1$ in the remaining case.

If $n \geq 3$, X being C.P., this degree is ≤ 0, and < 0 on at least one component of X : in fact $\deg \omega_X \geq 2$.

Let n be equal to 2, and let x, y belong to 2 different components:

then deg $F|_{X_i} \le 0$, and deg $F = 4 - 2\,p\,(X)$; therefore 2.6 fails if and only if $p\,(X) = 2$ and $\mathcal{O}_X\,(x+y) = \omega_X$; but then X consists of 2 components with deg $\omega_{X|X_i} = 1$.

It is then easy to see that either X consists of 2 elliptic tails, or X consists of two $\mathbf{P}^1$'s X_1, X_2 with $X_1 \cdot X_2 = 3$.

The former case though gives a contradiction, since then x should be a singular point of X ($\omega^{-1}_{X|X_1} = \mathcal{I}_{X_2} \otimes \mathcal{O}_{X_1}$), in the latter case we get a curve of genus 2.

If x, y belong to the same component Z, either $Z = X$ and $p\,(X) = 2$, or deg $(\omega_{X|Z}) = 1$; in this case every section of $F|_{X-Z}$ is identically zero, so we can apply Proposition A of [2], namely the following result

(2.7) Let L be an invertible sheaf on a curve X, s a non zero section of $H^0\,(X,L)$ such that s is identically 0 $(s \equiv 0)$ on $Y \subset X$, $s \not\equiv 0$ on any component of $Z = X-Y$: then $Y \cdot Z \le \deg L|_Z$

to obtain that Z is an elliptic tail.

<u>ii:</u> Let $\pi : \tilde{X} \to X$ be the normalization at x. Then $\mathrm{Hom}_{\mathcal{O}_X}\,(\mathfrak{m}_x\,\mathfrak{m}_y, \omega_X^{1-n})$ embeds in $H^0\,(\tilde{X}, L)$, where L is the invertible sheaf $\pi^*\,\omega_X^{1-n} \otimes \mathfrak{m}_y^{-1}$.

Clearly deg $L|_{\tilde{X}_i} < 0$ for every component $\tilde{X}_i$ of $\tilde{X}$, provided $n \ge 3$; if $n = 2$ this degree is ≤ 0 and $H^0\,(\tilde{X},L)$ can be non zero only if the irreducible component $\tilde{Y}$ of $\tilde{X}$ containing $\pi^{-1}(y)$ is a connected component of $\tilde{X}$, and deg $\pi^*\,\omega_{X|\tilde{Y}} = 1$. But then $H^0\,(\tilde{Y}, L|_{\tilde{Y}}) \neq 0$ implies that $L|_{\tilde{Y}}$ is trivial, hence $\tilde{C}\,\omega_{\tilde{Y}}\,(-y)$ is trivial: by a degree argument $p(\tilde{Y}) \le 1$, and actually $p\,(\tilde{Y}) = 0$ since $y \neq x$. Then deg $\tilde{C} = -3$ and $\pi: \tilde{Y} \to Y$ must be an isomorphism. Then $Y \cap (X-Y) = \{x\}$, $Y \cdot (X-Y) = 3$, therefore, from the fact that $|\omega_X^2|$ has no base points, either $H^0\,(X, \omega_X^2)$ maps onto $H^0\,(Y, \omega_X^2|_Y)$, so that x and y are separated by the bicanonical map, or this map restricted to Y is a double cover of $\mathbb{P}^1$, hence we go back to case i).

<u>iii</u>: Let $\pi\colon \tilde{X} \to X$ be the normalization at both x,y. Then $\mathrm{Hom}_{\mathcal{O}_X}(\mathfrak{m}_x \mathfrak{m}_y , \omega_X^{1-m})$ embeds in $H^0(\tilde{X}, \pi^* \omega_X^{1-n})$ which is 0 for $n \geq 2$ since X is C.P.

<u>iv</u>: Let $\pi\colon \tilde{X} \to X$ be the normalization at x. By lemma 2.2 $\mathrm{Hom}(\mathfrak{m}^2_x, \omega_X^{1-m})$ is a subspace of $H^0(\tilde{X},\mathcal{L})$, where $\mathcal{L}$ is the invertible sheaf $\pi^* \omega_X^{1-m} \otimes M^{-1}$.

Let $\tilde{Y}$ be a connected component of $\tilde{X}$, $\tilde{W}$ an irreducible component of $\tilde{Y}$. If $W = \pi(\tilde{W}) \not\ni x$, then $\deg \mathcal{L}|_{\tilde{W}} < 0$; if $W \ni x$, and x is smooth for W, $\deg \mathcal{L}|_{\tilde{W}} = 1 + (1-m) \deg \omega_X|_W$: this degree is then < 0 for $m \geq 3$, and, for $m = 2$, it is ≤ 0. If equality holds, $\deg \omega_X|_W = 1$, hence either $p(W) = 0$ or W is an elliptic tail (apply 1.12 to $\tilde{W} \subset \tilde{Y}$).

If x is a singular point of W, let C' be the conductor of $\mathcal{O}_{\tilde{W}}$ in $\mathcal{O}_W$, $\tilde{C}' = \pi^{-1}(C')$. We can write $\mathcal{L}$ as $\omega_{\tilde{Y}}^{-1} \otimes \pi^*(\omega_X^{2-m}) \otimes (\tilde{C}^{-1} \otimes M)^{-1}$, and let $d = \deg \omega_{\tilde{Y}}|_{\tilde{W}}$, $t = \deg (\tilde{C}^{-1} \otimes M)|_{\tilde{W}}$.

Since $C' \subset \mathfrak{M}_{x,W}$, $\tilde{C}' \subset M|_{\tilde{W}}$, $\dim \mathcal{O}_{\tilde{W}}/_{\tilde{C}} \leq 2\,\delta_W$, by 1.13 we conclude that $t \geq W(X-W)_x$. Hence either $(X-W) \not\ni x$, or $t \geq 2$ (in fact $W(X-W)_x = 1$ $\Rightarrow x$ is a node for X, but then, W being singular at x, $(X-W) \not\ni x$). In any case $\deg \mathcal{L}|_{\tilde{W}} \leq -t-d$, and if $t = 0$, then $C = M$, hence x is either a node or a cusp for W (and for X).

Consider the case when $d \geq 0$: then $\deg \mathcal{L}|_{\tilde{W}} \leq 0$, equality holding iff $d = t = 0$, $m = 2$ $(X-W \not\ni x)$.

Then, by 1.11, either $\tilde{W} = \tilde{Y}$ is elliptic, or $p(\tilde{W}) = 0$. In the former case $X = W$ is of genus 2, in the latter there exists a component

$\tilde{Z}$ of $\tilde{Y}$ on which $\deg \mathcal{L}|_{\tilde{Z}} < 0$, so that $H^0(\tilde{Y}, \mathcal{L}) = 0$.

If $d = -1$, $p(\tilde{W}) = 0$, $\tilde{W} \neq \tilde{Y}$, hence either $\deg \mathcal{L}|_{\tilde{W}} < 0$, or there exists $\tilde{Z} \subset \tilde{Y}$ with $\deg \mathcal{L}|_{\tilde{Z}} < 0$: if $\deg \mathcal{L}|_{\tilde{W}} = 0$, $H^0(\tilde{Y}, \mathcal{L}) = 0$, if $\deg \mathcal{L}|_{\tilde{W}} > 0$, then $t = 0$ and therefore W is an elliptic tail.

Finally, in the case when $d = -2$, $\tilde{W} = \tilde{Y} = \mathbb{P}^1$.

Assume that $(X-W) \not\ni x$, i.e. $W = X$; then, since $p(X) \geq 3$, $\delta_x \geq 3$. But then $C \subset \mathcal{M} \subset M \subset \pi_* \mathcal{O}_{\tilde{W}}$ and all the inclusions are strict ($\mathcal{M}$ is not an ideal in $\pi_* \mathcal{O}_{\tilde{W}}$), and since $\dim \mathcal{M}/C = \delta_x - 1$, $t = \dim M/C \geq 3$, and we are done.

If, on the other hand, $(X-W) \ni x$ and $t = 2$, by our previous argument $W(X-W)_x = 2$, $M = \tilde{C}'$, hence x is either a node or a cusp for W, which is thus Gorenstein with $\omega_W \cong \mathcal{O}_W$.

In this case, though $\deg \mathcal{L}|_{\tilde{W}} = 0$ for $m = 2$, we prove that, unless $C \not\subset \mathcal{M}^2_x$, any section of $\mathcal{H}om(\mathcal{M}^2_{x,X}, \omega_X^{-1})$ is $\equiv 0$ on W. Let $Z = X - W$, and consider the standard exact sequence

$$0 \to \mathcal{O}_X \to \mathcal{O}_W \oplus \mathcal{O}_Z \to \mathcal{O}_{Z \cap W} \to 0$$

Since $\mathcal{O}_{W \cap Z}$ has length 2, by Nakayama's lemma $\mathcal{M}^2_x \otimes \mathcal{O}_{W \cap Z} = 0$, and, if we set $\mathcal{M} = \mathcal{M}_x$, we have an isomorphism $\mathcal{M}^2 \mathcal{O}_X \cong \mathcal{M}^2 \mathcal{O}_W \oplus \mathcal{M}^2 \mathcal{O}_Z$: in fact $\mathcal{M}^2 \mathcal{O}_X$ injects into $\mathcal{M}^2 \mathcal{O}_W \oplus \mathcal{M}^2 \mathcal{O}_Z$, and clearly the projection on each factor is surjective; however $\mathcal{M}^2 \mathcal{O}_W = \pi_*(\tilde{C}|_{\tilde{W}})$ is contained in $C = \pi_*(\tilde{C})$, hence $\mathcal{M}^2 \mathcal{O}_W \subset \mathcal{M}^2 \mathcal{O}_X$ (and our assertion is thus proven) unless exists $f \in C - \mathcal{M}^2$ (this cannot hold if conjecture 3.23 is true).

Tensoring by ω_X^2, we get

$$H^1(X, \mathcal{M}^2 \omega_X^2) \cong H^1(W, \mathcal{M}^2 \mathcal{O}_W \otimes \omega_X^2) \oplus H^1(Z, \mathcal{M}^2 \mathcal{O}_Z \otimes \omega_X^2).$$

Use now R.R. duality on W, Z, respectively: this vector space is dual to $\mathrm{Hom}(\mathcal{M}^2 \mathcal{O}_W, \omega_X^{-2} \otimes \omega_W) \oplus \mathrm{Hom}(\mathcal{M}^2 \mathcal{O}_Z, \omega_X^{-2} \otimes \omega_Z)$.

We want to prove that the first summand is 0 : but here ω_W is trivial, hence this vector space embeds into $H^o(\tilde{W}, M^{-1}\pi^*\omega_X^{-2}|_{\tilde{W}}) = 0$ (these are the sections of a line bundle of degree -2).

Thus, given a section of $\operatorname{Hom}(\mathcal{M}^2_x, \omega_X^{-1})$, we have proven that it is 0 on any connected component $\tilde{Y}$ of $\tilde{X}$, except possibly if all irreducible components $\tilde{W}_i$ of $\tilde{Y}$ satisfy the following conditions (i=1,...r):

a) $x \in W_i$ and x is smooth for W_i

b) $p(W_i) = 0$, $W_i(X-W_i) = 3$.

It is easy to see that $r \leq 2$, so that either $W \cap (X-W) = \{x\}$, or W_1 intersects W_2 transversally at a node y of X.

In the latter case, if $X = Y$, then $p(X) = 2$, otherwise $2 = W_1(X-W_1)_x =$ = (by [6]) $= (W_1 \cdot W_2)_x + W_1(X-Y)_x$, hence $p(Y) = 1$, $Y(X-Y) = 2$ and we can repeat the argument given above.

We are left out with the case $W = \mathbb{P}^1$, $W \cap (X-W) = \{x\}$, $W \cdot (X-W) = 3$, i.e. b).

v: To end the proof of theorem C, let's prove the first statement. Namely, let W be an elliptic tail, and let x be the node of X such that $W \cap (X-W) = \{x\}$.

Then $\omega_X|_W = \mathcal{O}_W(x)$, and, since x is not a base point for the bicanonical map of X, $H^o(X, \omega_X)$ restricts onto $H^o(W, \mathcal{O}_W(2x))$, which has dimension 2, by R.R.

Therefore, under the bicanonical map of X, W is a double cover of $\mathbb{P}^1$.

If, finally, $p(X) = 2$, let s_o, s_1 be a basis of $H^o(X, \omega_X)$. Assume that $s_o \cdot s_1 \equiv 0$: then, if X_i is the largest curve $\subset X$ s.t s_i does not vanish identically on any component of X_i, $\dim X_o \cap X_1 = 0$, and, by 2.7, since X is connected, $1 \leq X_i \cdot (X-X_i) \leq \deg \omega_X|_{X_i}$.

But $p(X) = 2 \Longrightarrow \deg \omega_X = 2$, hence $\deg \omega_{X|X_i} = 1$, so X_o, X_1 are elliptic tails; since moreover X is C.P. $X = X_o \cup X_1$.

If, on the other hand, $\forall s, \sigma \in H^o(X, \omega_X)$, $s \cdot \sigma \not\equiv 0$, $s_o^2, s_o s_1, s_1^2$ constitute a basis for $H^o(X, \omega_X^2)$, and the bicanonical map of X is a double cover of a smooth conic in $\mathbb{P}^2$.

Q.E.D.

§ 3. The Canonical Map

Throughout this section we will continue to assume that X is a complete reduced C.P. Gorenstein curve.

In order to discuss the behavior of the canonical map of X, we need some definitions.

Remark 3.1: If X is not 2 connected according to def. 1.3, there exists, by 1.10, a Z such that $Z \cap (X-Z) = x$, and x is a node for X: such an x is called a disconnecting node.

Definition 3.2: An irreducible component Y of X with $p(Y) = 0$ is said to be a loosely connected rational tail (L.C.R.T.) if $Y (X-Y)$ equals the number of connected components of $X-Y$.

Remark 3.3: If Y is a L.C.R.T., Y intersects $(X-Y)$ in disconnecting nodes. The next result gives necessary and sufficient conditions in order that the canonical map be a morphism.

Theorem D: If X is C.P. the base locus of $|\omega_X|$ consists exactly of the L.C.R.T.'s and of the disconnecting nodes. So $|\omega_X|$ is free from base points if and only if X is 2-connected.

Proof: Consider the exact sequence

$$0 \to \mathfrak{m}_x \, \omega_X \to \omega_X \to k_x \to 0 .$$

Then x is a base point if $H^1(\mathfrak{m}_x \, \omega_X) \to H^1(\omega_X) \to 0$ is not an isomorphism, i.e. if and only if $h^1(\mathfrak{m}_x \, \omega_X) = 2$, i.e. $\dim \operatorname{Hom}(\mathfrak{m}_x, \mathcal{O}_X) = 2$.

Step I: Assume now x to be a singular point of X. By 2.1, if $\hat{X}$ is the blow-up of X at x, $\operatorname{Hom}(\mathfrak{m}_x, \mathcal{O}_X) \to H^0(\hat{X}, \mathcal{O}_{\hat{X}})$, hence if x is a base point $\hat{X}$ is not connected, and a fortiori $\tilde{X}$ is not connected ($\pi: \tilde{X} \to X$ being the normalization at x).

Let $\pi^{-1}(x) = \{p_1, \ldots p_k\}$, and let t_i be a uniformizing parameter at p_i, m_i the multiplicity of $\tilde{C}$ at p_i, D the divisor $\sum_{i=1}^{k} m_i p_i$ on $\tilde{X}$.

Let W be the vector space $\omega_{\tilde{X}}(D)/\omega_{\tilde{X}}$: every element η of W can be written in an unique way as $\sum_{i=1}^{k} \sum_{j_i=1}^{m_i} a_{i,j_i} \left(dt_i / t_i^{j_i} \right)$, and the dimension of W is $2\delta_x$.

W contains the vector subspace $V = \{\eta \mid \forall f \in \mathcal{O}_x \mid \sum_{i=1}^{k} \operatorname{Res}_{p_i}(f \cdot \eta) = 0\}$, of dimension δ_x, and a local section of ω_X around x is a local section of $\omega_{\tilde{X}}(D)$ around the p_i's such that its image in W belongs to V.

Moreover a local generator ω_X lifts to a differential with pole of order exactly m_i at each p_i, (so that $a_{i,m_i} \neq 0$), and via this choice, one can identify $\mathcal{O}_{X/C}$ with V.

Let us denote by U the image of $H^0(\omega_X)$ in $\mathcal{O}_{X/C}$: by what we just said, we can view U as a subspace of V.

Consider now the exact sequence

$$(3.4) \quad 0 \to H^0(\omega_{\tilde{X}}) \to H^0(\omega_{\tilde{X}}(D)) \to \underset{\displaystyle = W}{\omega_{\tilde{X}}(D)/\omega_{\tilde{X}}} \xrightarrow{d} H^1(\omega_{\tilde{X}}) \to 0$$

Let K be the kernel of d. It is clear then that $U = K \cap V$, and proving that x is not a base point amounts to proving that there exists a vector in U with some $a_{i,m_i} \neq 0$. We have described the linear equations which define V: we claim now that the elements of K are those who satisfy the following equations

$$(3.5) \qquad \sum_{p_i \in \tilde{Y}} a_{i,1} = 0, \text{ for each connected component } \tilde{Y} \text{ of } \tilde{X}.$$

In fact (3.4) is the direct sum of the exact sequences on each $\tilde{Y}$, and then there is a canonical isomorphism of $H^1(\omega_{\tilde{Y}})$ with k, given by the trace map.

Take $w = \sum_{i=1}^{k} w_i$, where $w_i = \sum_{1 \leq j_i \leq m_i} a_{i,j_i} (dt_{i/t_i^{j_i}})$. Then $w|_{\tilde{Y}} =$

$= \sum_{p_i \in \tilde{Y}} w_i$, and if you take A_i an open set in $\tilde{Y}$ where the above expression for w_i gives a section of $\omega_{\tilde{Y}}(D)$ (assume $A_i \not\ni p_j$ for $i \neq j$), and set $A_o = \tilde{Y} - \pi^{-1}(x)$, $d|_{\tilde{Y}}(w)$ is given by the cocycle $(w_i - w_j)$ on $A_i \cap A_j$.

Following the same argument given in [7], page 248, we see that we get the zero element in $H^1(\tilde{Y}, \omega_{\tilde{Y}})$ if $\sum \operatorname{Res} w_i = 0$, i.e. if (3.5) holds.

By 1.14 $m_i \geq 2$ unless x is a node; assume then that x is not a node.

We can decompose $W = W' \oplus W''$, where W' is the span of the (dt_{i/t_i})'s, W'' is the span of the (dt_{i/t_i^j}), for $j \geq 2$.

Consider the equations defining V: if $f = 1$ we get the equation $\sum_{i=1}^{k} a_{i,1} = 0$, if $f \in \mathfrak{m}_x$ we get an equation involving only the $(a_{i,j})$'s with $j \geq 2$.

We can therefore conclude that $K = K' \oplus W''$, $V = V' \oplus V''$, $V' \supset K'$, hence $U = K \cap V = K' \oplus V''$.

Since there is a vector in V" with $a_{i,m_i} \neq 0$, we infer that x is not a base point for $|\omega_X|$. If, instead, x is a node which disconnects, we have $\pi^{-1}(x) = \{p_1, p_2\}$, $m_i = 1$ and must be, for vectors in K, $a_{1,1} = a_{2,1} = 0$, so x is a base point of $|\omega_X|$.

Step II: Let x be a smooth point of X, Z the irreducible component of X to which x belongs. Let L be the line bundle $\mathcal{O}_X(x)$: if x is a base point $h^0(X,L) = 2$, in particular $h^0(Z, \mathcal{O}_Z(x)) = 2$, hence $p(Z) = 0$. Let y be a point of $Z \cap (X-Z)$, and s a non zero section of L vanishing at y: then s vanishes identically exactly on a curve W which is a union of connected components of (X-Z). By (2.7) $Z \cdot W = 1$, so y is a disconnecting node and W is connected. Therefore Z is a L.C.R.T.

Conversely, if Z is a L.C.R.T., $y_i, \ldots y_r$ are the disconnecting nodes belonging to Z, $\omega_X|_Z = \omega_Z(y_1 + \ldots + y_r)$, but since every section of ω_X vanishes at the y_i's by Step I, every section of ω_X vanishes identically on Z.

Q.E.D.

Remark 3.6: If X is C.P. and connected, but not 2-connected, one can take the normalization of X at the disconnecting nodes, to obtain $\pi : Y \to X$, where $Y = \bigcup_{i=1}^{m} Y_i$ consists of (say) m connected components. It is straight forward to verify that $H^0(X, \omega_X) \cong H^0(Y, \omega_Y) = \bigoplus_{i=1}^{m} H^0(Y_i, \omega_{Y_i})$, and that the Y_i's are 2-connected curves; in other words the rational canonical map $|\omega_X|$ consists of π^{-1} followed by the canonical morphisms of the Y_i, whose images span projective subspaces in a skew position. Therefore it is not restrictive to consider only the canonical map of a 2-connected curve.

In the rest of the paragraph we are going to examine necessary and sufficient conditions in order that the canonical map be an embedding, and we shall often start with some example just to explain some definitions and results. The first question is whether $|\omega_X|$ is injective, and we have the following

Definition 3.7: X is strongly connected if there do not exist two nodes x, y of X such that $X - \{x\} - \{y\}$ is disconnected. In particular, if X is 3-connected, then X is strongly connected.

Theorem E: If X is 2-connected, C.P., and the canonical map is injective, then X is strongly connected. More precisely, if x, y are two singular points of X, they have the same image under $|\omega_X|$ if and only if x, y are nodes and $X - \{x\} - \{y\}$ is disconnected.

Proof: If x, y are not nodes, we can repeat the argument given in Step I of Theorem D. Namely, let $\tilde{X}$ be the normalization of X at x, y, $\pi: \tilde{X} \to X$, $\tilde{C} = \mathcal{O}_{\tilde{X}}(-D_1 - D_2)$, where D_1, D_2 are effective divisors with supp $(D_1) = \pi^{-1}(x)$, supp $(D_2) = \pi^{-1}(y)$.

Let $W_1 = \omega_{\tilde{X}}(D_1 + D_2)/\omega_{\tilde{X}}(D_2) \supset \mathcal{O}_{x,X}/C = V_1$, and let W_2, V_2 be defined in an analogous way.

Again we can decompose V_i as $V_i' \oplus V_i''$, and if $V = V_1 \oplus V_2$, U is the image of $H^0(\omega_X)$ in $W = \omega_{\tilde{X}}(D_1 + D_2)/\omega_{\tilde{X}}$, $U = K \cap V = K' \oplus V''$, where $K' \subset V'$ and $U \supset V'' = V_1'' \oplus V_2''$, so that there exists sections of ω_X vanishing at x but not at y, and conversely.

Assume instead that x is a node, and let X' be the normalization of X at x: then $C = \mathcal{M}_{x,X'}$, therefore $H^0(X, \omega_X \mathcal{M}_{x,X}) = H^0(\omega_{X'})$, and x, y have the same image under $|\omega_X|$ if and only if y is a base point for $|\omega_{X'}|$. The result follows then immediately from Theorem D.

Q.E.D.

Remark 3.8: Let X be 2-connected, C.P., but not strongly connected, x_1 a node of X such that the normalization X' of X at x is not 1-connected, but has $(r-1)$ disconnecting nodes $x_2, \ldots x_r$. Then, if Φ is the canonical map, $\Phi(x_i)$, for $i=1,..r$, is a fixed point p of $\Phi(X) = C$.

Let $\tilde{X}$ be the normalization of X at the x_i's: then the effect of projecting C from p is the same than to consider the canonical map of $\tilde{X}$. Therefore we obtain easily in this way examples where the canonical map is not injective, though being birational.

We are now going to discuss hyperelliptic curves, i.e. those for which $|\omega_X|$ is not birational.

Definition 3.9: X is hyperelliptic if there exist 2 smooth points x, y (possibly $x = y$) such that $H^0(\mathcal{O}_X(x+y)) = 2$.

Proposition 3.10: Let X be 2-connected. X is hyperelliptic if and only if $|\omega_X|$ is not birational, and also if and only if two smooth points have the same image, or $|\omega_X|$ is not an embedding at a smooth point.

Proof: The second part follows immediately by the exact sequence

$$0 \to H^0(\omega_X(-x-y)) \to H^0(\omega_X) \to \omega_X / \mathfrak{m}_x \mathfrak{m}_y \omega_X \to$$

$$\to H^1(\omega_X(-x-y)) \to H^1(\omega_X) \to 0 \text{, since the dual space}$$

of $H^1(\omega_X(-x-y))$ is $H^0(\mathcal{O}_X(x+y))$.

For the first part, notice that $H^0(\mathcal{O}_X(x+y))$ defines a morphism $f: X \to \mathbf{P}^1$, so that, for a general $p \in \mathbb{P}^1$, $f^{-1}(p)$ consists of two smooth points x', y', which have the same image under $|\omega_X|$.

Q.E.D.

<u>Example 3.11</u>: Let G be a cubic surface in $\mathbb{P}^3$ with an ordinary quadratic singularity at P, and containing exactly 6 lines through P. Let π_1, π_2 be two planes tangent to G at P, and such that $\pi_i \cdot G = Y_i$ is an irreducible cubic curve. Let Q be the point where Y_1, Y_2 intersect transversally ($\pi_1 \cdot \pi_2 \cdot G = 2P + Q$), and blow up $\mathbb{P}^3$ at Q.

The strict transform X of $Y = Y_1 \cup Y_2$ is a genus 3 curve, and it is easy to see that the canonical map of X is given by projection with center Q, hence the canonical map has as its image two lines in $\mathbb{P}^2$, and has degree 2 on each component.

<u>Example 3.12</u>: Notice first that the union of 2 conics in $\mathbb{P}^2$ is canonically embedded. Here the cross ratio of the 4 points in a conic through them determines uniquely the conic in the pencil determined by the 4 base points. Consider now, on $\mathbb{P}^1 \times \mathbb{P}^1$, two irreducible curves of type $(1,n)$, $(1,m)$ respectively: they have $p = 0$, and intersect in $(n+m)$ points (possibly infinitely near). It is easy to see that the canonical map is induced by the complete linear system $|\mathcal{O}_{\mathbb{P}^1 \times \mathbb{P}^1}(0, n+m-2)|$, hence it is given by the projection on the second factor of $\mathbb{P}^1 \times \mathbb{P}^1$, followed by the embedding of $\mathbb{P}^1$ as a rational normal curve of degree $(n+m-2)$. Here the cross ratio of any 4- of the $(n+m)$ points is the same on both curves.

<u>Remark 3.13</u>: Let Y be an irreducible hyperelliptic curve: thus there exists a morphism $f : Y \to \mathbb{P}^1$ of degree 2. Then f is finite, and exists n such that $f_* \mathcal{O}_Y = \mathcal{O}_{\mathbb{P}^1} \oplus \mathcal{O}_{\mathbb{P}^1}(-n)$. In particular Y is a

divisor in a smooth surface (a line bundle over $\mathbb{P}^1$), hence Y is Gorenstein, and has at most double point as singularities.

Proposition 3.14: Let X be 2-connected, and let x, y be smooth points of X such that $h^o(\mathcal{O}_X(x+y)) = 2$.

Then either

a) x, y belong to 2 different components Y_1, Y_2 with $p(Y_i) = 0$, and such that for every connected component Z_j of $X-Y_1-Y_2$ $Z_j \cdot Y_i = 1$,

or

b) x, y belong to an irreducible hyperelliptic curve Y such that, for each connected component Z of $X-Y$ the invertible sheaf $(\mathcal{I}_Z \otimes \mathcal{O}_Y)^{-1}$ is isomorphic to the hyperelliptic sheaf $\mathcal{O}_Y(x+y)$.

Proof: Let L be the invertible sheaf $\mathcal{O}_X(x+y)$. By assumption $h^o(L) = 2$ hence, for every $z \neq x, y$, $h^o(\mathfrak{m}_{z,X}L) = 1$.

In case a), pick up z either on $Y_1 \cap Y_2$ or, if $Y_1 \cap Y_2 = \phi$, in a connected component Z of $X-Y_1-Y_2$ such that $Z \cap Y_i \neq \phi$.

Let s be a non zero section of $H^o(X, L)$ vanishing at z: since $L|_{X-Y_1-Y_2}$ is trivial, s vanishes at some point of Y_i else than x, or y.

The section s cannot vanish identically on any of the Y_i's : in fact it cannot vanish on both Y_1 and Y_2, so assume $s|_{Y_1} \equiv 0$, $s|_{Y_2} \not\equiv 0$.

Let W be the union of connected components of $X - Y_2$ where $s \equiv 0$: by 2.7 $Y_2 \; W \leq 1$, hence X would not be 2-connected $(W(X-W) \leq 1)$, a contradiction.

Therefore the restriction map $H^o(X, L) \to H^o(Y_i, L|_{Y_i})$ is an isomorphism and $p(Y_i) = 0$.

By the same argument, for each connected component Z of $X - Y_1 - Y_2$, $Z \cdot Y_i \leq 1$, and since X is 2-connected $Z\ (Y_1 \cup Y_2) \geq 2$, hence $Z \cdot Y_i = 1$. In case b), if Z is a connected component of $X - Y$, $Z \cdot Y \leq 2$ by 2.7, so equality holds by 2-connectedness. Moreover $h^0\ (Y, L|_Y) = 2$, so Y is hyperelliptic; hence Y is Gorenstein and by 1.12 $\mathcal{J}_Z \otimes \mathcal{O}_Y$ is invertible, of degree -2.

Since there exists a non zero section s of $H^0\ (X, L)$ such that $s \in H^0\ (\mathcal{J}_Z\ L)$, we get an inclusion $0 \to \mathcal{O}_Y \to \mathcal{J}_Z L \otimes \mathcal{O}_Y$, and the cokernel of the map is a skyscraper sheaf of length 0, therefore

$$\mathcal{J}_Z \otimes \mathcal{O}_Y \cong L^{-1}|_Y\ .$$

Q.E.D.

To sharpen the result of the last proposition, and also prove a converse statement characterizing hyperelliptic reducible curves, it is convenient to have a digression on cross-ratios and rational normal curves (cf. 3.12).

<u>Definition 3.15</u>: An n-tuple of points on a smooth curve X consists of the following data: and ideal sheaf $\mathcal{J}$ of $\mathcal{O}_X$ such that length $\mathcal{O}_{X/\mathcal{J}} = n$, together with isomorphisms α_i for each $p_i \in$ supp $(\mathcal{O}_{X/\mathcal{J}})$, of $(\mathcal{O}_{X/\mathcal{J}}) \to k\,[t]/_{(t^{m_i})}$, (where $m_i =$ = length $(\mathcal{O}_{X,p_{i}/\mathcal{J}})$).

<u>Definition 3.16</u>: Two n-tuples of points on $\mathbb{P}^1$, $(\mathcal{J},\alpha_i)$, $(\mathcal{J}',\alpha'_i)$ are said to have the same cross ratios if there exists an automorphism of $\mathbb{P}^1$ such that $g^*\ (\mathcal{J}) = \mathcal{J}'$, and $g^* : \mathcal{O}_{\mathbb{P}^1/\mathcal{J}} \to \mathcal{O}_{\mathbb{P}^1/\mathcal{J}'}$ is such that $\alpha_i' \circ g^* = \alpha_i$.

Let now Y_1, Y_2 be two smooth rational curves of the same degree d in

$\mathbb{P}^N$, and $X = Y_1 \cup Y_2$. Then they have an n-tuple of points in common if length $\mathcal{O}_{Y_1 \cap Y_2} = n$, because, if $p_i \in Y_1 \cap Y_2$, any isomorphism α_i of $\mathcal{O}_{Y_1 \cap Y_2, p_i}$ to $k[t]/(t^{m_i})$ induces an n-tuple of points on Y_1 and Y_2. It makes thereforesense to say that Y_1 and Y_2 have n points in common with the same cross ratios.

Lemma 3.17: Let Y_1, Y_2 be two rational normal curves of degree d in $\mathbb{P}^N$ with n points in common. If $n \geq d + 3$, or $n = d + 2$ and the 2 n-tuples have the same cross-ratios, then $Y_1 = Y_2$.

Proof: Let's prove the result by induction on d.

For $d = 2$ the result is elementary and well known (one has only to remark that the hypothesis implies that the 2 conics lie in the same plane). So assume the theorem to be true for d-1.

Take a point $p \in Y_1 \cap Y_2$ and consider the projection $g: \mathbb{P}^N \to \mathbb{P}^{N-1}$ with centre p. Then $g(Y_1)$, $g(Y_2)$ satisfy the hypotheses of the lemma, hence $g(Y_1) = g(Y_2)$, so Y_1, Y_2 are contained in the cone Γ_{d-1} over the rational normal curve of degree (d-1) (in particular $\Gamma_{d-1} \subset \mathbb{P}^d \subset \mathbb{P}^N$). Consider $F = F_{d-1}$ the rational ruled surface obtained by blowing up p in $\Gamma = \Gamma_{d-1}$, $\pi: F \to \Gamma$ being the resolution of singularities. In Pic(F), let f be the class of a fibre, e the class of the exceptional divisor E; $\pi^*(\mathcal{O}_{\mathbb{P}^d}(1)) \equiv (d-1) f + e$, and we have $e^2 = -d + 1$, $e f = 1$, $f^2 = 0$. Let's denote by Y'_i the proper transform of Y_i. Since $Y'_i \cdot ((d-1) f + e) = d$, $Y'_i \cdot e = 1$, $Y'_i \equiv d f + e$, hence $Y'_1 \cdot Y'_2 = d + 1$. However $Y'_1 \cdot Y'_2 \geq n - 1$, hence it is a contradiction to assume $Y_1 \neq Y_2$ if $n \geq d + 3$. If, on the other hand, $n = d + 2$ and the cross-ratios of the n-tuple of points is the same, then the equations of Y'_1, Y'_2 induce the same element of $H^0(E, \mathcal{O}_F(Y'_i)/\mathcal{O}_F(Y_i - mE))$, where m is the length of $\mathcal{O}_{Y_1 \cap Y_2}$ at p: hence again $Y'_1 \cdot Y'_2 \geq n$ and we have a contradiction.

Q.E.D.

<u>Definition 3.18</u>: An honestly hyperelliptic curve is a 2-connected Gorenstein curve Y with a finite morphism $f: Y \to \mathbb{P}^1$ of degree 2.

<u>Theorem F</u>: A C.P., 2-connected Gorenstein curve X is hyperelliptic if and only if X contains an honestly hyperelliptic curve Y with an hyperelliptic invertible sheaf L on Y such that, for each connected component Z of $X-Y$, $(\mathcal{J}_Z \otimes \mathcal{O}_Y)^{-1}$ is isomorphic to L. Moreover, if $f : Y \to \mathbb{P}^1$ is the morphism associated to $H^0(Y, L)$, then the canonical map Φ_1 maps Y to a rational normal curve and factors through f. If Y is not irreducible, the above condition is equivalent to : $Y = Y_1 \cup Y_2$ with $p(Y_i) = 0$, and s.t. for every connected component Z_j of $X - Y$, $Z_j \cdot Y_i = 1$, and moreover, if we set $P_{ij} = Y_i \cap Z_j$, the n-tuples $(Y_1 \cap Y_2, P_{1j})$, $(Y_2 \cap Y_1, P_{2j})$ have the same cross-ratios.

<u>Proof</u>: Assume X to be hyperelliptic, and let x, y be to smooth points which have the same image under Φ_1. Following the arguments of 3.14., the invertible sheaf $L' = \mathcal{O}_X(x+y)$ defines a morphism $f': X \to \mathbb{P}^1$, which is non constant on a curve Y ($= Y_1 \cup Y_2$ in case a)). $f = f'|_Y$ makes Y a honestly hyperelliptic curve, and it is easy to see that $\Phi_{1|Y}$ factors through f.

Since $L' = f'^*(\mathcal{O}_{\mathbb{P}^1}(1))$ and a connected component Z of $X - Y$ is f'^{-1} (point), the argument of 3.14 gives $L'|_Y = L \cong (\mathcal{J}_Z \otimes \mathcal{O}_Y)^{-1}$.

Conversely, we claim that we can extend L to an invertible sheaf L' on X such that $L'|_Y = L$, $L'|_{X-Y} \cong \mathcal{O}_{X-Y}$: in fact we have the exact sequence

$$0 \to \mathcal{O}_{Y \cup Z} \to \mathcal{O}_Z \oplus \mathcal{O}_Y \to \mathcal{O}_{Y \cap Z} \to 0,$$

so choose a section s_Z of L not vanishing at $Y \cap Z$ and identify it with $1 \in H^0(\mathcal{O}_Z)$; in this way we have defined an $\mathcal{O}_{Y \cup Z}$ invertible sheaf, so, repeating the operation for each Z, we obtain L with the desired property.

Clearly $H^0(X, L) = H^0(Y, L')$, therefore $\Phi_{1|Y}$ factors through f and X is hyperellptic.

Let $Y \cdot (X - Y) = 2k$, $p = p(Y)$: then $\omega_{X|Y} = f^*(\mathcal{O}_{\mathbb{P}^1}(k + p - 1))$.

It remains to prove that, via f^*, $H^0(\mathbb{P}^1, \mathcal{O}_{\mathbb{P}^1}(k + p - 1)) =$
$= H^0(X, \omega_X)|_Y$.

Observe that $X - Y = Z_1 \cup \ldots \cup Z_k$, and that by R.R. $H^0(Z_j, \omega_{X|Z_j}) =$
$= p_j + 1$, where $p_j = p(Z_j)$.

In other words, $H^0(Z_j, \omega_X|_{Z_j}) \to \omega_X \otimes \mathcal{O}_{Y \cap Z_j}$ has a 1-dimensional image giving a local generator of ω_X at the points of $Y \cap Z_j$.

From the exact sequence

$$0 \to H^0(X, \omega_X) \to H^0(Y, \omega_X|_Y) \oplus H^0, (X-Y\ \omega_{X|X-Y}) \to$$

$$\to H^0(Y \cap (X-Y), \omega_X \otimes \mathcal{O}_{Y \cap (X-Y)}) \to H^1(X, \omega_X) \to 0$$

if follows easily that $H^0(\omega_X)|_Y$ has dimension $k + p$.

The last assertion follows by definition 3.16: in fact there exist isomorphisms $g_i : Y_i \to \mathbb{P}^1$ and ideals $\mathcal{J}, \mathcal{J}_i$ on $\mathbb{P}^1$ such that

a) $g_1^*(\mathcal{J}) = \mathcal{O}_{Y_1} \otimes \mathcal{J}_{Y_2}$ (resp. $g_2^*(\mathcal{J}) = \ldots.$), $g_i^*(\mathcal{J}_j) = \mathcal{O}_{Y_i} \otimes \mathcal{J}_{Z_j}$

b) $(g_1^*)^{-1}(g_2^*)$ induces the identity on $\mathcal{O}_{\mathbb{P}^1/\mathcal{J}}$, and on $\mathcal{O}_{\mathbb{P}^1/\mathcal{J}_j}$

when $P_{1j} = P_{2j}$ hence g_1, g_2 glue to give a finite morphism $f : Y \to \mathbb{P}^1$ of degree 2 such that $f^*(\mathcal{J}_j) = \mathcal{J}_{Z_j} \otimes \mathcal{O}_Y$.

Q.E.D.

Let us assume, for the rest of the paragraph, that X is not hyperelliptic and that X is strongly connected: then the canonical map Φ_1 is a birational morphism, is an embedding at smooth points, and separates pairs of smooth points as well as pairs of singular points. The next proposition ensures that Φ_1 is an injective morphism.

Proposition 3.19: Let X be C.P. and 2-connected. Then if x is a singular point, and y is a smooth point, $\Phi_1(x) \neq \Phi_1(y)$.

Proof: Let $\tilde{X}$ be the normalization at x. Since $H^0(\omega_{\tilde{X}}) = H^0(C\omega_X)$, if $H^0(\mathfrak{m}_x \mathfrak{m}_y \omega_X) = H^0(\mathfrak{m}_x \omega_X)$, y would be a base point for $|\omega_{\tilde{X}}|$. If Γ is the component of X containing y, it follows that $\tilde{\Gamma}$ is a L.C.R.T. (or contained in a negative tail).

Take now the normalization of $\tilde{X}$ at the points of $\tilde{\Gamma} \cap (\tilde{X} - \tilde{\Gamma})$, to get $\pi: \bar{X} \to X, \bar{\Gamma} \to \Gamma$, and let $\pi^*(\omega_X) = \omega_{\bar{X}}(\bar{D})$.
Choose t an affine coordinate on $\bar{\Gamma} \simeq \mathbf{P}^1$ such that $p_1, \dots p_k$ are the coordinates of the points in $\bar{\Gamma} \cap \pi^{-1}(x)$, $q_1, \dots q_s$ the ones of the points lying over $\tilde{\Gamma} \cap (\tilde{X} - \tilde{\Gamma})$ (they do not lie in $\pi^{-1}(x)$!).
Let also $z_1, \dots z_r$ be the points in $(\bar{X} - \bar{\Gamma}) \cap \pi^{-1}(x)$, t_i be a local coordinate at z_i, let $u_1, \dots u_s$ be the points of $\bar{X} - \bar{\Gamma}$ lying over $\tilde{\Gamma} \cap (\tilde{X} - \tilde{\Gamma})$, and let τ_h be a local coordinate at u_h.
The multiplicity of $\bar{D}$ at u_h, q_h, is one, and let m_i be the multiplicity of $\bar{D}$ at p_i, n_j the multiplicity of $\bar{D}$ at z_j.
Consider the usual exact sequence:

$$0 \to H^0(\omega_{\bar{X}}) \to H^0(\omega_{\bar{X}}(\bar{D})) \to W = \omega_{\bar{X}}(\bar{D})/\omega_{\bar{X}} \xrightarrow{\partial}$$

$$\xrightarrow{\partial} H^1(\omega_{\bar{X}}) \to 0$$

An element η in W can be written in the form

$$\sum_{\substack{i=1 \\ 1\leq j\leq m_i}}^{k} a_{ij} \; dt\,(t-p_i)^{-j} + \sum_{h=1}^{s} a_h \, dt\,(t-q_h)^{-1} +$$

$$+ \sum_{h=1}^{s} b_h \, d\tau_h \, (\tau_h)^{-1} + \sum_{\substack{e=1 \\ 1\leq n\leq n_e}}^{r} c_{e,n} \, dt_e \, (t_e)^{-n}$$

We can clearly assume that $m_i \geq 2$ for each i (otherwise x would be a node and X would not be either C.P. or 2-connected). Remark also that $H^0(\omega_{\bar{X}})\mid_{\bar{\Gamma}} = 0$. An element $\eta \in W$ is in the image U of $H^0(\omega_X)$ if and only if it belongs to the intersection of two subspaces, K' and V.

K' is defined by the equations of $K = \operatorname{Ker} \partial$ plus the local equations given by the nodes in $(\tilde{X} - \tilde{\Gamma}) \cap \tilde{\Gamma}$: since $\bar{X}$ has at least $s+1$ connected components, there is given, for each $h = 1,\ldots s$, a subset I_h of $\{1,\ldots r\}$, and also are given subsets $J_{h'}$, $h'=1,\ldots p$, such that the I_h's and $J_{h'}$'s give a partition of $\{1,\ldots r\}$ and K' is defined by the following equations

$$\sum_{i=1}^{k} a_{i1} + \sum_{h=1}^{s} a_h = 0 \quad ,$$

$$a_h + b_h = 0 \quad (h=1,..s), \quad b_h + \sum_{e\in I_h} c_{e,1} = 0 \, (h=1,..s)$$

$$\sum_{e\in J_{h'}} c_{e,1} = 0 \quad (h'=1,..p).$$

The subspace V is defined by the equations $\sum \operatorname{Res}(f\eta) = 0$ for $f \in \mathfrak{m}_x$, and the variables $a_{i1}, a_h, b_h, c_{e,1}$ do not appear in these equations. The conditions that η vanishes at x is given by any of the equations $a_{i\,m_i} = 0$ (these are all equivalent to each other modulo the equations defining V) : again the above mentioned variables do not appear.

An easy computation around $y = \infty$ gives that η vanishes at y if and only if

$$(\#) \qquad \sum_{i=1}^{k} a_{i1}\, p_i + \sum_{h=1}^{s} a_h\, q_h + \sum_{i=1}^{k} a_{i2} = 0.$$

If then $H^o(\mathcal{M}_x \mathcal{M}_y \omega_X) = H^o(\mathcal{M}_x \omega_X)$, the equation (#) should be a linear combination of the equations defining K', V, and of the equations $a_{im_i} = 0$.

By looking at the coefficient of the a_{i1}'s, we get that all the p_i's should be equal. This however is possible only if $k = 1$, and then we can assume $p_1 = 0$.

Look now at the coefficient of a_{12} : a_{12} appears in the equations defining V if and only if Γ is smooth at x.

If Γ is singular at x, a_{12} appears in (#) with coefficient 1, and it has non zero coefficient in the other equations only if $m_1 = 2$.

But then, by 1.13., $X - \Gamma \not\ni x$ and x is an ordinary cusp. In this case we have a contradiction again since either $X = \Gamma$ has genus 1, or X is not 2-connected. Assume finally Γ to be smooth at x.

Restrict the linear form (#) to the subspace where

$$a_{1j} = 0 \text{ for } j \geq 2, \quad c_{e,j} = 0 \text{ for } j \geq 2.$$

Then the linear form $\sum_{h=1}^{s} a_h q_h$ should be a linear combination of the linear forms

$$a_{11} + \sum_{h=1} a_h, \quad a_h + b_h \quad (h=1,..s), \quad b_h + \sum_{e \in I_h} c_{e,1}$$

$$(h=1,...s), \quad \sum_{e \in J_{h'}} c_{e,1} \quad (h' = 1,..p).$$

This is however easily seen to be impossible.

Q.E.D.

<u>Example 3.20</u>: Let X_1, X_2 be two smooth non hyperelliptic curves meeting in a point x such that $(X_1 \cdot X_2)_x = 2$ (a tacnode), set $X = X_1 \cup X_2$. Then, if s is a section of ω_X vanishing at x, $s|_{X_i} \in H^o(X_i, \omega_{X_i}(2x))$ hence it vanishes to second order on X_i at x.

This shows that X is non hyperelliptic, and the canonical map is not an embedding at x.

This motivates the following

Definition 3.21: A (C.P.) Gorenstein curve X is said to be very strongly connected if

a) X is strongly connected

b) there does not exist a decomposition $X = X_1 \cup X_2$ where $X_1 \cap X_2$ is a single point.

Proposition 3.22: Assume that X is very strongly connected and that x is a double point (i.e. formally isomorphic to the plane singularity $y^2 - x^k = 0$). Then the canonical map is not an embedding at x if and only if X is hyperelliptic with $f\colon X \to \mathbf{P}^1$ not constant on the components of X passing through x.

Proof: Let's prove first the "if" part of the statement.

Let Γ be the union of the components passing through x (they are at most 2).

Then by Theorem F the restriction to Γ of the canonical map of X factors through f, hence is not an embedding at x.

Conversely, let $\tilde{X}$ be the normalization of X at x: then by our hypothesis $\tilde{X}$ is connected.

Moreover, $\dim \operatorname{Hom} (\mathcal{m}^2_x, \mathcal{O}_X) = 2$, and, by lemma 2.2, $\tilde{X}$ is such that $h^0 (\tilde{X}, M^{-1}) \geq 2$, hence $\tilde{X}$ is hyperelliptic, with $L = M^{-1}$ as hyperelliptic bundle.

We have in fact that if $h^0 (\tilde{X}, L) = 3$, then $\tilde{X}_1, \tilde{X}_2$ must be negative tails (if they were L.C.R.T. X would not be 2-connected), and then X

is clearly hyperelliptic with the desired properties (according to Theorem F). So we can assume $h^o(\tilde{X}, L) = 2$, and that $p(\tilde{X}) \geq 1$. Observe that there exists an integer r s.t. $\pi^*(\omega_X) = \omega_{\tilde{X}} \otimes L^r$ (in fact $r = [k/2]$).

Let σ be a section of L vanishing on $\pi^{-1}(x)$, τ a section of L vanishing at two smooth points p,q of $\tilde{\Gamma}$ but not on $\pi^{-1}(x)$, η a section of $H^o(\omega_{\tilde{X}})$ such that $\eta|_{\tilde{\Gamma}}=$ a power of τ.

By the exact sequence

$$0 \to H^o(\omega_{\tilde{X}}) \to H^o(\omega_X) \to \omega_{X}/C\omega_X \to 0$$

we see that $\eta\tau^h\sigma^{r-h}$ $(h = 1,\dots r)$ are sections of $\pi^*(\omega_X)$ which, in $\omega_{\tilde{X}}/M^{-r}\omega_{\tilde{X}}$, give a basis of $\omega_X/C\omega_X$.

Therefore the pull-back of sections of ω_X, when restricted to $\tilde{\Gamma}$, are linear combinations of $\tau^{h'}\sigma^{h''}$, hence p and q have the same image under Φ_1 and X is hyperelliptic.

Q.E.D.

<u>Theorem G</u>: Let X be very strongly connected, not hyperelliptic, and such that for each singular point x of X where $C \not\subset \mathfrak{m}_x^2$, x is a double point. Then the canonical map Φ_1 is an embedding.

<u>Remark 3.23</u>: The hypotheses in Theorem G would only be that X be very strongly connected and not hyperelliptic if the following conjecture were true: any Gorenstein singular point x where $C \not\subset \mathfrak{m}_x^2$ is a double point. This is obvious if $\dim \mathfrak{m}_x/\mathfrak{m}_x^2 = 2$ and we shall later give a proof of this fact when the singularity is unibranch, i.e. formally irreducible. Also case a) of Theorem C would be vacuous if the conjecture were true.

<u>Proof of Theorem G</u>: In view of 3.19, 3.22, we are only left to prove that Φ_1 is an embedding at a singular point x where $\mathfrak{m}_x^2 \supset C$.

Consider the exact sequence

$$0 \to H^0(C\omega_X) \to H^0(\omega_X) \to \omega_{X/C\omega_X} \to$$

$$\to H^1(C\omega_X) \to H^1(\omega_X) \to 0 : \text{ since } H^1(C\omega_X) =$$

$$= H^1(\pi_* \omega_{\tilde{X}}) = H^1(\omega_{\tilde{X}}), \text{ and } \tilde{X} \text{ is connected, the}$$

restriction homomorphism $H^0(\omega_X) \to \omega_{X/C\omega_X}$ is surjective. Also $\mathcal{O}_{X/C}$ surjects onto $\mathcal{O}_{X/\mathfrak{m}_x^2}$, and we are done.

Q.E.D.

<u>Proposition 3.26</u>: Let (X,x) be a reduced Gorenstein unibranch singularity (i.e. if $\pi: \tilde{X} \to X$ is the normalization at x, $\pi^{-1}(x)$ is a single point p). If $C \not\subset \mathfrak{m}_x^2$, then x is a double point.

<u>Proof</u>: Let t be a uniformizing parameter in $\tilde{\mathcal{O}} = \mathcal{O}_{\tilde{X},p}$, and let M be the semigroup $M = \{\text{ord}_t f \mid f \in \mathfrak{m}_x\}$. Notice that $M \not\ni 1$, and we can assume that $M \not\ni 2$, otherwise then x is a double point. Take a function $g \in C - \mathfrak{m}_x^2$ such that $m = \text{ord}_t g$ is maximum (observe that $\mathfrak{m}_x^2 \supset C^2 = (t^{4\delta})$, so $\text{ord}_t g \leq 4\delta$).

Then we claim that $m \notin M + M$. Otherwise if $m = m_1 + m_2$, $m_i = \text{ord}_t f_i$, $f_i \in \mathfrak{m}_X$, there would exist a constant λ such that $\text{ord}_t (g - \lambda f_1 f_2) > m$, but, since $(g - \lambda f_1 f_2) \in C - \mathfrak{m}^2$ (in fact $\text{ord}_t f \geq 2\delta \Leftrightarrow f \in C$), this contradicts the maximality of m.

If $1 \leq n_1, \ldots n_r \notin M$, then the r-dimensional subspace $\sum_{i=1}^{r} \lambda_i t^{n_i}$ intersects $\mathcal{O} \subset \tilde{\mathcal{O}}$ only in 0, and since $\dim \tilde{\mathcal{O}}/\mathcal{O} = \delta$, it follows easily that $\delta \geq \text{card}(\mathbb{N} - M)$ (actually one has equality). Moreover, by definition of C, $(2\delta - 1) \notin M$. Consider now the $[m/2]$ pairs $\{1, m-1\}$, $\{2, m-2\}, \ldots$: since $m \notin M + M$ at least one element for each pair does not belong to M, therefore $\delta \geq [m/2]$, hence $1 + 2\delta \geq m$. Since $m \geq 2\delta$, either $m = 2\delta$, but then we have noticed that $1, 2\delta - 1 \notin M$, or $m = 2\delta + 1$ but then, $2, 2\delta - 1 \notin M$.

Q.E.D

§ 4. Some Remarks on the Non-Reduced Case

Let C be a smooth curve of genus g, L, N line bundles on it, and consider C as the zero section of $V = L \oplus N$ (a smooth non complete threefold with a projection $p: V \to C$).

The sheaves of sections of L, N, pull back, via p, to invertible sheaves $\mathcal{L}, \mathcal{N}$ on V. The normal sheaf to C in V is clearly $(\mathcal{L} \oplus \mathcal{N}) \otimes \mathcal{O}_C$, hence $\omega_{V|C} = \omega_C \otimes \mathcal{L}^{-1} \otimes \mathcal{N}^{-1}$. Let $X \to V$ be the curve (locally complete intersection) defined by the ideal I_X spanned by $\mathcal{L}^{-2} + \mathcal{N}^{-2}$ (here $\mathcal{L}^{-1}, \mathcal{N}^{-1}$ are viewed as given by linear forms on the fibres of $p: V \to C$). Therefore $X_{red} = C$, and the conormal sheaf to X in V is given by $(\mathcal{L}^{-2} \oplus \mathcal{N}^{-2}) \otimes \mathcal{O}_X$, hence $\omega_X = (\omega_V \otimes \mathcal{L}^2 \otimes \mathcal{N}^2)|_X = (\omega_C \otimes \mathcal{L} \otimes \mathcal{N})|_X$ (again here ω_C stands for the pull back via p). Let d be the degree of $(\omega_C \otimes \mathcal{L} \otimes \mathcal{N})_C$, then it follows that $\deg \omega_X = 4d$, for instance since we have the exact sequence

$$0 \to (\mathcal{L}^{-1} \oplus \mathcal{N}^{-1} \oplus \mathcal{L}^{-1}\mathcal{N}^{-1})|_C \to \mathcal{O}_X \to \mathcal{O}_C \to 0$$

and we can tensor it by ω_X^n to compute $\chi(\omega_X^n)$. Since C is a subscheme of X, if $|\omega_X^n|$ is free from base points or embeds, the analogous statement must hold true a fortiori for $|\omega_X^n|_C|$.

But, if $\mathcal{L}, \mathcal{N}$ are chosen to be general in Pic (C), one needs $nd \geq 2g$ (respectively $nd \geq 2g + 1$). This is a lower bound on n which however depends on deg $(\omega_{X|C})$, compared to deg $(\omega_C) = 2g - 2$, i.e. on the negativity of the normal bundle to C.

But consider now the following (non closed) double point of X :

a tangent vector sticking out of a point $x \in C$ in the direction of N, together with x.

In other words, we consider the subscheme of X defined by the ideal $J = p^* (\mathcal{M}_{x, C}) + \mathcal{L}^{-1}$.

This double point is not embedded if $H^o (J \omega_X{}^n)$ has codimension ≤ 1 in $H^o (\omega_X{}^n)$.

This clearly happens if x is a base point of $|\omega_X{}^n|_C|$, and, in the other case, one can consider the following diagram:

$$\begin{array}{ccccccc} & & 0 & & 0 & & 0 \\ & & \downarrow & & \downarrow & & \downarrow \\ 0 \to \omega_X{}^n \otimes & & (\mathcal{L}^{-1} \oplus \mathcal{L}^{-1}\mathcal{N}^{-1} \oplus \mathcal{M}_x \mathcal{N}^{-1})|_C & \to & J\,\omega_X{}^n & \to \mathcal{M}_x & (\omega_X{}^n|_C) \to 0 \\ & & \downarrow \gamma & & \downarrow \alpha & & \downarrow \beta \\ 0 \to \omega_X{}^n \otimes & & (\mathcal{L}^{-1} \oplus \mathcal{L}^{-1}\mathcal{N}^{-1} \oplus \mathcal{N}^{-1})|_C & \to & \omega_X{}^n & \to & (\omega_{X|C}^n) \to 0 \end{array}$$

Assume that γ induces an isomorphism of H^o's: then

cod $\mathrm{Im}\, H^o(\alpha) \leq$ cod $\mathrm{Im}\, H^o(\beta) \leq 1$.

$H^o(\gamma)$ is clearly an isomorphism iff

$$0 \to H^o (\omega_X{}^n \mathcal{N}^{-1} \otimes \mathcal{M}_{x,C}) \to H^o (\omega_X{}^n \mathcal{N}^{-1} \otimes \mathcal{O}_C)$$

is an isomorphism.

A sufficient condition for this to hold is that

$$H^o(C, \omega_X{}^n \mathcal{N}^{-1} \otimes \mathcal{O}_C) = 0, \text{ e.g. if}$$

$$\deg \mathcal{N}|_C > nd = n\,(2g - 2 + \deg \mathcal{L}|_C + \deg \mathcal{N}|_C).$$

This condition means that if $\deg \mathcal{N}|_C = m$, $\deg \mathcal{L}|_C = e$, m must be very positive, e very negative, and yet the degree of the normal bundle to C, $\delta = m + e$, can be positive. In fact the above inequality is then

$$m > n\,(2g - 2) + n\,\delta\,.$$

The conclusion is that the hypothesis of the normal bundle to C being positive still does not give any lower bound for n in order that $|\omega_X{}^n|$ be an embedding.

If X is a curve lying on a smooth surface, then one can define, according to Franchetta and Ramanujam (see [9], [5], [1]) a notion of numerical m-connectedness for X : it would be interesting to extend this notion for a Gorenstein curve, and to see whether some conditions of this kind can give some results of the type of Theorems A, B.

References

[1] Bombieri, E. - Canonical Models of Surfaces of General Type, Publ. Math. I.H.E. S.42 (1973), 171-219.

[2] Bombieri, E. - Catanese, F. - The Tricanonical Map of a Surface with $K^2 = 2$, $p_g = 0$, "C. P. Ramanujam - A Tribute", Stud. in Math. 8, Tata Inst. Bombay (1978), Springer, 279-290.

[3] Catanese, F. - Le Applicazioni Pluricanoniche di una Curva Riducibile Giacente su una Superficie, Publ. Ist. Mat. "L. Tonelli" (1979), Pisa.

[4] Deligne, P. - Mumford, D. - The Irreducibility of the Space of Curves of Given Genus, Publ. Math. I.H.E.S. 36 (1969) 75-110.

[5] Enriques, F. - "Le Superficie Algebriche", Zanichelli, Bologna (1949).

[6] Harris, J. - Thetacharacteristics on Singular Curves, preprint

[7] Hartshorne, R. - "Algebraic Geometry", Springer GTM 52 (1977).

[8] Mumford, D. - "Lectures on Curves on an Algebraic Surface", Annals of Math. Studies, 59, Princeton (1966).

[9] Ramanujam, C. P. - Remarks on the Kodaira Vanishing Theorem, J. Ind. Math. Soc. 36 (1972), 41-51.

[10] Sakai, F. - Canonical Models of Complements of Stable Curves, Int. Symp. Alg. Geom. Kyoto, Iwanami Shoten, (1977), 643-661.

[11] Serre, J. P. - "Groupes Algebriques et Corps de Classes", Act. Sc. et Ind. 1264, Hermann, Paris (1959).

Fabrizio Catanese, Ist. Mat. "L. Tonelli" Università di Pisa (Via Buonarroti 2, 56100 PISA), and Institute for Advanced Study, (Princeton, NJ 08540), a member of G.N.S.A.G.A. of C.N.R.

The author is indebted to the Institute Mittag-Leffler and to the Institute for Advanced Study for their warm hospitality and was partly suppored by N.S.F. grant MCS 81-03365 during his stay at the I.A.S.

POSITIVITY AND EXCESS INTERSECTION

by

William Fulton and Robert Lazarsfeld

§0. Introduction.

Consider a variety M, and a projective local complete intersection

$$X \subseteq M$$

of pure codimension e. Then for any subvariety $Y \subseteq M$ of dimension $k \geq e$, the intersection class

$$X \cdot Y \in A_{k-e}(X)$$

is defined up to rational equivalence on X. One of the most basic facts of intersection theory is that if Y meets X, and does so *properly*, then $X \cdot Y$ is non-zero and in fact has positive degree with respect to any projective embedding of X. On the other hand, if the intersection of X and Y is improper, then $X \cdot Y$ may be zero or of negative degree. Our purpose here is to give some conditions on X to guarantee the non-negativity or positivity of the intersection class in the case of possibly *excess* intersection. These conditions take the form of hypotheses on the normal bundle $N_{X/M}$ to X in M, the theme being that positivity of the vector bundle $N_{X/M}$ forces the positivity of $X \cdot Y$ provided only that Y meets X. We give several simple applications and related results, including a lower bound for the multiplicity of a proper intersection, generalizing a classical result for curves on a surface.

§1. Excess intersections with positive normal bundle.

We deal with a variety M - not necessarily smooth or complete - and a local complete intersection $X \subseteq M$ of pure codimension e, which we assume to be projective. Denote by N the normal bundle to X in M, and let L be a fixed ample line bundle on X. We are interested in intersecting X with a subvariety $Y \subseteq M$ of dimension $k \geq e$.

Theorem 1. (A). *If $S^m(N)$ is generated by its global sections for some $m > 0$, then*

$$\deg_L(X \cdot Y) \geq 0.$$

(B). *If $S^m(N) \otimes \check{L}$ is generated by its global sections for some $m > 0$, then*

$$\deg_L(X \cdot Y) \geq \frac{1}{m^{\dim(X)}} \cdot \deg_L(X \cap Y).$$

(For an ℓ-dimensional cycle or cycle class α on X, $\deg_L(\alpha)$ denotes the degree of the zero-dimensional class $c_1(L)^{\ell} \cap \alpha$. In (B), $\deg_L(X \cap Y)$ is the sum of the L-degrees of the irreducible components of $X \cap Y$, taken with their reduced structures.) The hypothesis in (B) is equivalent to the assumption that the normal bundle N is *ample* in the sense of Hartshorne [H1], and the proof will show that in fact a somewhat better inequality holds.

Before proceeding, we record several simple applications:

Corollary 1. *In the situation of the theorem, if $S^m(N)$ is generated by its global sections for some $m > 0$, then X is numerically effective in the sense that*

$$(*) \qquad \deg(X \cdot Y) \geq 0$$

for any subvariety $Y \subseteq M$ of pure dimension $e = \operatorname{codim}(X)$. If moreover N is ample, then strict inequality holds in (*) provided that Y is numerically equivalent to an effective cycle whose support meets X. ■

Corollary 2. *Let $V_1, \ldots, V_r \subseteq \mathbf{P}^n$ be subvarieties of degrees $d_1, \ldots, d_r$. Then*

$$\deg(V_1 \cap \ldots \cap V_r) \leq d_1 \cdot \ldots \cdot d_r \ .$$

This was originally proved by R. MacPherson and the first author. As before, the left-hand side denotes the sum of the degrees of the irreducible components of $V_1 \cap \ldots \cap V_r$ with their reduced structures.

Proof. By passing to a larger projective space, and to cones over the V_i, we may assume that $\Sigma \dim(V_i) \geq (r-1)n$. Let $M = \mathbf{P}^n \times \ldots \times \mathbf{P}^n$ (r times), and let $X = \mathbf{P}^n \subseteq M$ be the diagonal. The hypotheses of statement (B) of the theorem are satisfied with $L = \mathcal{O}_{\mathbf{P}^n}(1)$ and $m = 1$. Taking $Y = V_1 \times \ldots \times V_r$, the corollary follows. ■

Exercise. Assuming $r = 2$, show that if equality holds in Corollary 2 then V_1 and V_2 lie in a linear subspace of $\mathbf{P}^n$ in which they meet properly.

Corollary 3. *In the setting of the theorem, suppose that M is acted on transitively by a connected algebraic group. If the normal bundle N to X in M is ample, then X meets any subvariety $Y \subseteq M$ of dimension* $\geq \operatorname{codim}(X)$.

Proof. The homogeneity of M implies that Y is algebraically equivalent to a subvariety Z which meets X, and $X \cdot Z \neq 0$ by the theorem. ■

This simplifies and extends somewhat a result of Lübke [L].

Remark. Corollary 3 is closely related to two conjectures of Hartshorne ([H2] III.4.4, III.4.5) concerning smooth subvarieties of a non-singular variety M in characteristic zero.

Conjecture A. If $X \subseteq M$ has an ample normal bundle, then some multiple of X moves (as a cycle) in a large algebraic family.

Conjecture B. If both $X \subseteq M$ and $Y \subseteq M$ have ample normal bundles, and if $\dim(X) + \dim(Y) \geq \dim(M)$, then X meets Y.

Observe that if one knew that some multiple of $[X]$ moved in an algebraic family large enough to cover M, then Conjecture B would follow from Theorem 1 as in the proof of Corollary 3. However, one has the following:

Counter-example to Conjecture A. Gieseker [Gi] has constructed an ample vector bundle E on $\mathbf{P}^2$ arising as a quotient of the form

$$(*) \qquad 0 \to \mathcal{O}_{\mathbf{P}^2}(-n)^2 \to \mathcal{O}_{\mathbf{P}^2}(-1)^4 \to E \to 0.$$

for suitable $n >> 0$. Take M to be the total space of E, and $X \subseteq M$ to be the zero-section. We claim that there are no projective surfaces $Y \subseteq M$ other than X itself, from which it follows that no multiple of X moves in any non-trivial algebraic family. Indeed, an embedding $Y \subseteq M$ distinct from the zero-section would give rise to a non-zero section of f^*E on the normalization $\tilde{Y}$ of Y, where f is the composition of the natural maps $\tilde{Y} \to Y \to X$. But $H^1(\tilde{Y}, f^*\mathcal{O}_{\mathbf{P}^2}(-n)) = 0$ by the Mumford-Ramanujam vanishing theorem, and hence $H^0(\tilde{Y}, f^*E) = 0$.

Conjecture B remains open, although it seems to us plausible that a counter-example may exist. The general picture that appears to emerge from ([Han], [Fa], [Go], [L], [FL1]) is that ampleness of normal bundles has global consequences for subvarieties of homogeneous spaces, but not necessarily in general.

There are two inputs to the proof of Theorem 1. The first, which is the essential feature of the intersection theory developed by R. MacPherson and the first author ([FM1], [FM2], [Fu]), consists in reducing to an infinitesimal problem. Specifically, in the situation of the theorem, consider the fibre square

$$\begin{array}{ccc} X \cap Y & \subseteq & Y \\ \cap\!\!| & & \cap\!\!| \\ X & \subseteq & M, \end{array}$$

and denote by C the normal cone of $X \cap Y$ in Y. Then C has pure dimension $k = \dim Y$, and sits naturally as a subscheme in the total space of the normal bundle $N = N_{X/M}$. One can intersect C with the zero section of N to obtain a well-defined rational equivalence class

$$z(C,N) \in A_{k-e}(X) .$$

($z(C,N)$ is actually defined in $A_{k-e}(X \cap Y)$.) We call this the cone class determined by C in N. The basic fact then is that

$$X \cdot Y = z(C,N)$$

(cf. [FM1], [Fu]). Theorem 1 now follows from a general positivity statement for the classes determined by cones in vector bundles satisfying hypotheses (A) or (B) of the theorem:

Theorem 2 (cf. [FL2]). *Let N be a vector bundle of rank e on a projective variety X, and let $C \subseteq N$ be an irreducible cone of dimension $k \geq e$. Let L be an ample line bundle on X.*

(A) *If $S^m(N)$ is generated by its global sections for some $m > 0$, then*

$$\deg_L(z(C,N)) \geq 0.$$

(B) *If $S^m(N) \otimes \check{L}$ is generated by its global sections for some $m > 0$, then*

$$\deg_L(z(C,N)) \geq \frac{1}{m^{\dim(\mathrm{Supp}\, C) + e-k}}\, s(C) \deg_L(\mathrm{Supp}\, C),$$

where $s(C)$ is the multiplicity of C along its zero-section.

We will prove the theorem under the stronger hypotheses:

(A') N is generated by its global sections.

(B') $N \otimes \check{L}$ is generated by its global sections.

The general case is treated by combining the proof below with the arguments in §2 of [FL2].

Proof. We may assume that $\mathrm{Supp}(C) = X$. If N is generated by its global sections, then a general section of N meets C properly or not at all. Therefore, $z(C,N)$ is represented by an effective (or zero) cycle, and this proves (A).

Turning to (B), after possibly replacing $C \subseteq N$ by $C \oplus L \subseteq N \oplus L$ --which leaves the cone class unchanged ([FL2§1])--we may assume that C maps to its support with fibre dimension ≥ 1. In this situation, one has the formula

$$z(C,N) = \pi_*(c_{e-1}(Q_{\mathbf{P}(N)}) \cap [\mathbf{P}(C)]),$$

where $Q_{\mathbf{P}(N)} = \pi^* N/\mathcal{O}_{\mathbf{P}(N)}(-1)$ is the rank $e-1$ universal quotient bundle on the projectivization $\pi : \mathbf{P}(N) \to X$. Thus

$$\deg_L(z(C,N)) = \int_{\mathbf{P}(C)} c_1(\pi^*L)^{k-e}\, c_{e-1}(Q_{\mathbf{P}(N)})$$

$$= \sum_{i=0}^{e-1} \int_{\mathbf{P}(C)} c_1(\pi^*L)^{k-e+i}\, c_{e-1-i}(Q_{\mathbf{P}(N)} \otimes \pi^*L^\vee) .$$

By the hypothesis (B'), $Q_{\mathbf{P}(N)} \otimes \pi^*L^\vee$ is generated by its global sections, and it follows that its Chern classes are represented by effective (or zero) cycles. Thus all the terms in the sum above are non-negative. Therefore, letting $n = \dim X = \dim \mathrm{Supp}(C)$, one has

$$\deg_L(z(C,E)) \geq \int_{\mathbf{P}(C)} c_1(\pi^*L)^n\, c_{k-1-n}(Q_{\mathbf{P}(N)} \otimes \pi^*L^\vee)$$

$$= \int_{\mathbf{P}(C)} c_1(\pi^*L)^n\, c_{k-1-n}(Q_{\mathbf{P}(N)})$$

$$= \deg\{c_1(L)^n \cap s_n(C,N)\} ,$$

where $s_n(C,N) \in A_n(X)$ is the n-dimensional Segre class of C. But $s_n(C,N) = s(C) \cdot [X]$, and the theorem follows. ∎

Remark. As a special case of statement (B) of Theorem 2, one finds that if N is ample, and if $C \subseteq N$ is a cone of pure dimension $e = \mathrm{rk}(N)$, then the cone class $z(C,N)$ has strictly positive degree. This was proved in [FL2], where it was used to determine all numerically positive polynomials in the Chern classes of an ample vector bundle.

§2. Intersection multiplicities.

If C and D are curves on a surface M, and $P \in C \cap D$ is an isolated point of intersection, then a classical formula of Max Noether expresses the intersection multiplicity $m_P(C \cdot D)$ of the given curves at P in terms of their proper transforms on the blow-up $\tilde{M}$ of M at P. Specifically, Noether's formula states that

$$m_P(C \cdot D) = e_P(C) \cdot e_P(D) + \sum_{Q \in E} m_Q(\tilde{C} \cdot \tilde{D}),$$

$\tilde{C}$ and $\tilde{D}$ being the proper transforms of C and D, where the sum on the right is taken over all points on the exceptional divisor $E \subseteq \tilde{M}$. In particular, since this sum is non-negative, one obtains the familiar lower bound for the intersection multiplicity. In this section we discuss a generalization of Noether's formula to higher dimensions. Unlike the situation for curves on a surface, it can happen in general that the proper transforms of the given varieties no longer meet properly in a neighborhood of the exceptional divisor. In this case, positivity results come into play in order to bound from below the contribution of this intersection.

Let $V_1,\dots,V_r$ be subvarieties of a smooth variety M, with $\Sigma\,\mathrm{codim}(V_i,M) = \dim(M)$. Assume that $V_1,\dots,V_r$ intersect properly at the point $P \in M$. As we are interested in local questions, we will suppose that P is the only point at which the V_i meet. Denote by $\tilde{M}$ the blow-up of M at P, so that the exceptional divisor E is a projective space, with

$$L = \mathcal{O}(-E)\,|_E$$

the (ample) hyperplane bundle. If $\tilde{V}_i \subseteq \tilde{M}$ is the blow-up of V_i at P, then $\cap \tilde{V}_i$ is contained in E, and hence $\tilde{V}_1 \cdot \ldots \cdot \tilde{V}_r$ is a well-defined rational equivalence class of dimension zero on E.

Theorem 3. *With the preceding notation,*

(A) $m_P(V_1 \cdot \ldots \cdot V_r) = \prod_{i=1}^{r} e_P(V_i) + \deg(\tilde{V}_1 \cdot \ldots \cdot \tilde{V}_r)$, *where* $e_P(V_i)$ *is the multiplicity of* V_i *at* P.

(B) $\deg(\tilde{V}_1 \cdot \ldots \cdot \tilde{V}_r) \geq \deg_L(\tilde{V}_1 \cap \ldots \cap \tilde{V}_r)$.

We will prove (A) and (B) when each V_i is a divisor on M. The proof of (A) in general uses the deformation to the normal bundle, as in [FM1]. For (B), one cannot apply Theorem 1 to the diagonal imbedding of $\tilde{M}$ in $\tilde{M} \times \ldots \times \tilde{M}$, since the normal bundle to this imbedding is not ample. Instead, one imbeds $\tilde{M}$ in the blow-up of $\tilde{M} \times \ldots \times \tilde{M}$ along $E \times \ldots \times E$, where the normal bundle is ample. We refer to [Fu] §12.4 for details.

Proof. Let $\pi : \tilde{M} \to M$ and $\eta : E \to P$ be the canonical maps. From the projection formula and the identification of $\mathcal{O}(E)\,|_E$ with $\mathcal{O}(-1)$, one has:

(i) $\eta_*(\pi^* V_1 \cdot \ldots \cdot \pi^* V_r) = V_1 \cdot \ldots \cdot V_r$

(ii) $\eta_*(\pi^* V_1 \cdot \ldots \cdot \pi^* V_j \cdot E^{r-j}) = 0 \qquad (1 \leq j < r)$

(iii) $\eta_*(E^r) = (-1)^{r-1}[P]$.

Let $m_i = e_P(V_i)$. Equivalently, one has an equation of divisors on $\tilde{M}$:

(iv) $\pi^* V_i = m_i E + \tilde{V}_i \qquad (1 \leq i \leq r)$.

By (i) - (iv) and bilinearity of intersection products,

$$\eta_*(\tilde{V}_1 \cdot \ldots \cdot \tilde{V}_r) = \eta_*((\pi^* V_1 - m_1 E) \cdot \ldots \cdot (\pi^* V_r - m_r E))$$
$$= V_1 \cdot \ldots \cdot V_r - m_1 \cdot \ldots \cdot m_r [P]$$

which proves (A). Shrinking M, we may assume each V_i is a principal Cartier divisor on M, so that, by (iv), $\mathcal{O}(\tilde{V}_i) = \mathcal{O}(-m_i E)$. The intersection class $\tilde{V}_1 \cdot \ldots \cdot \tilde{V}_r$ may be constructed from the fibre square

$$\begin{array}{ccc} \cap \tilde{V}_i & \subseteq & \tilde{M} \\ \cap\! | & & \cap\! | \\ \tilde{V}_1 \times \ldots \times \tilde{V}_r & \subseteq & \tilde{M} \times \ldots \times \tilde{M} . \end{array}$$

The normal bundle to $\tilde{V}_1 \times \ldots \times \tilde{V}_r$ in $\tilde{M} \times \ldots \times \tilde{M}$ restricts to $\bigoplus_{i=1}^{r} L^{\otimes m_i}$ on $\cap \tilde{V}_i \subseteq E$. Theorem 2(B) then applies, as in the proof of Theorem 1, to show that

$$\deg \tilde{V}_1 \cdot \ldots \cdot \tilde{V}_r \geq \deg_L(\tilde{V}_1 \cap \ldots \cap \tilde{V}_r) . \quad ■$$

<u>Remark</u>. When the V_i are Cartier divisors, the theorem holds even if M is singular. In fact, the preceding proof shows that if m_i are any positive integers such that

$$\pi^* V_i = m_i E + W_i$$

for some effective Cartier divisors W_i on $\tilde{M}$, then

$$m_P(V_1 \cdot \ldots \cdot V_r) = \prod_{i=1}^{r} m_i \cdot e_P(M) + \deg(W_1 \cdot \ldots \cdot W_r)$$

with $\deg(W_1 \cdot \ldots \cdot W_r) \geq \deg_L(W_1 \cap \ldots \cap W_r)$. In place of (iii) one uses the equation $\eta_*(E^r) = (-1)^{r-1} e_P(M) \cdot [P]$.

REFERENCES

[Fa] G. Faltings, Formale Geometrie und homogene Räume, Inv. math. 64(1981), 123-165.

[Fu] W. Fulton, Intersection Theory, forthcoming.

[FL1] W. Fulton and R. Lazarsfeld, Connectedness and its applications in algebraic geometry, in Algebraic Geometry Proceedings: Chicago, 1980, Lect. Notes in Math. 862(1981), 26-92.

[FL2] W. Fulton and R. Lazarsfeld, The numerical positivity of ample vector bundles, to appear.

[FM1] W. Fulton and R. MacPherson, Intersecting cycles on an algebraic variety, in Real and Complex Singularities, Oslo, 1976, Sitjhoff and Noordhoff (1978), 179-197.

[FM2] W. Fulton and R. MacPherson, Defining algebraic intersections, Algebraic geometry Proceedings: Tromsϕ, 1977, Lect. Notes in Math. 687(1978), 1-30.

[Gi] D. Gieseker, P-ample bundles and their Chern classes, Nagoya Math. J. 43(1971), 91-116.

[Go] N. Goldstein, Ampleness in complex homogeneous spaces and a second Lefschetz theorem, preprint.

[Han] J. Hansen, Connectedness theorems in algebraic geometry, Proceedings of the 18th Scandanavian Congress of Mathematics, 1980, Birkhäuser Boston (1981).

[H1] R. Hartshorne, Ample vector bundles, Publ. Math. I.H.E.S. 29(1966), 63-94.

[H2] R. Hartshorne, Ample Subvarieties of Algebraic Varieties, Lect. Notes in Math. 156(1970).

[L] M. Lübke, Beweis einer Vermutung von Hartshorne für den Fall homogener Mannigfaltigkeiten, J. für die reine und ang. Math. 316(1980), 215-220.

William Fulton: Institut des Hautes Études Scientifiques, Institute for Advanced Study, and Brown University. Partially supported by the Guggenheim Foundation, the NSF, and the Alfred P. Sloan Foundation.

Robert Lazarsfeld: Institute for Advanced Study and Harvard University. Partially supported by an AMS Postdoctoral Research Fellowship.

NOTES ON THE EVOLUTION OF COMPLETE CORRELATIONS

Dan LAKSOV

1. The problem of correlations between two d-dimensional spaces in an n-dimensional space can be stated in the following way:

"In a projective space [n] of dimension n, determine all the pairs of linear subspaces S_d and S'_d of dimension d together with a correlation between them, such that S_d and S'_d satisfy given Schubert conditions $a_0, a_1, \ldots, a_d$ and $a'_0, a'_1, \ldots, a'_d$ respectively, and the correlation satisfies a composite condition $\mu_0^{n_0}, \mu_1^{n_2}, \ldots, \mu_{d-1}^{n_{d-1}}$ where

$$\Sigma_{i=0}^{d} (a_i + a'_i) + d = \Sigma_{i=0}^{d} n_i$$

and where μ_i is the condition that the two i-dimensional spaces in which S_d and S'_d in general meet two fixed linear spaces R_{n-d+i} and R'_{n-d+i} respectively, are conjugate under the correlation."

As usual a linear space S_d is said to satisfy a Schubert condition imposed by a flag of subspaces $A_0 \subset A_1 \subset \ldots \subset A_d$ of [n], where $\dim A_i = a_i$, if $\dim(S_d \cap A_i) \geqq i$ for $i = 0,1,\ldots,d$.

A (non-degenerate) correlation between two spaces S_d and S'_d is a linear, non-degenerate, map between S_d and the dual space S'^*_d of S'_d. The subspaces Q_i and Q'_i of S_d and S'_d respectively are said to be conjugate under the correlation if the (d-i-1)-dimensional subspace of S'_d which correspond to Q_i under the correlation, intersect the space Q'_i.

The relation between the numbers a_i, a'_i, d and n_i is chosen such that the number of pairs of correlated spaces satisfying the given conditions ought to be finite according to a counting of constants.

The fundamental work on the problem of correlations was done by Hirst. He first gave an appropriate formulation of the problem in analogy with the similar problem for conics treated by Chasles and generalizing problems treated by Sturm on the correlations on lines. Then in order to solve the problem in the case $d = n = 2$ he introduced the notion of exceptional or degenerate correlations and showed that they play the same fundamental role in the study of correlations as the degenerate conics do in Chasles theory of conics. His methods immediately became part of the folklore of the time and various cases in the line plane and space were solved by Hirst [6], [7], [8], [9], [10], [11], Loria [12], Sturm [30], [31], [32] and Visalli [35], [36], [37]. Later H. Schubert [14], [15], [16], [17], [18], [19], [20], [21] concerned himself with the problem in higher dimensional spaces. As we shall indicate, the general directions for such generalizations are already given in Hirsts work. The main contribution of Schubert was to set up a procedure for solving the problem of correlations by induction and of giving explicit formulas in the cases $n_0 = n_1 = \dots = n_{d-4} = 0$, the most beautiful of which is the number

$$\begin{vmatrix} \binom{a_0+a_0'}{a_0}, & \binom{a_1+a_0'}{a_1} & \cdots & \binom{a_d+a_0'}{a_d} \\ \binom{a_0+a_1'}{a_0} & \vdots & & \vdots \\ \vdots & & & \\ \binom{a_0+a_d'}{a_0} & & \cdots & \binom{a_d+a_d'}{a_d} \end{vmatrix}$$

for the correlations satisfying the conditions

$$(a_0,a_1,\dots,a_d)(a_0',a_1',\dots,a_d')\mu_{d-1}^{n_{d-1}} \text{ with}$$

$$\Sigma_{i=0}^{d}(a_i+a_i') + d = n_{d-1},$$

that is for the numbers of pairs of spaces S_d and S_d' with a correlation between them such that the intersection of S_d with n_{d-1} fixed hyperplanes in general position are pairwise conjugate to the intersection of S_d' with another set of n_{d-1} hyperplanes in general position. Schubert only gave the method for proving formulas of the above type and also proved similar formulas for quadrics. The formulas for correlations was proved and generalized by Giambelli by the methods indicated by Hirst and Schubert, but with more refined combinatorial techniques.

Later authors have been interested in the geometric aspects of the complete correlations rather than in their enumerative properties. This study was initiated by F. Severi and taken up by Semple and Tyrrell. Semple indicated how to construct a parameter space for both complete quadrics and correlations and how to study in detail the properties of these spaces. He also indicated how to prove the fundamental degeneration relations. Semples program was completed by Tyrrell.

Our interest in this subject comes from the observation of A. Lascoux that the formulas obtained by Giambelli are similar to formulas obtained for the characteristic classes of symmetric products of vector bundles. The combinatorial techniques for handling such classes is old and their applications to questions pertaining to geometry has had a very successful renaissance during the last years. Lascoux suggested that it would be most interesting to see how these combinatorial techniques apply to the problem of correlations. We shall return to the general problem in a forthcoming article with Lascoux.

In the present article we shall concentrate on the geometry behind the combinatorial formalism. In order to appreciate the enumerative theory, it is of utmost importance to understand the geometry of the degenerate correlations and the conditions imposed upon them. The main problem in the interpretation of the enumerative formulas is that in every linear family of non-degenerate correlations there are limiting positions lying in the codimension one locus of degenerate linear maps and several "essentially different" families have the same limiting positions. Hence in every linear family of correlations there will be degenerate ones that must be taken into account and to do this the limiting positions must be separated. This is exactly the purpose of the degenerate correlations. Hirst shows in his work how the degenerate correlations can be considered as codimension one subfamilies of the space of all complete correlations and showed in a beautiful geometric manner how the degenerate correlations appear in the linear families of correlations and how they contribute to the enumerative formulas. We shall, in the following, expose the original geometric methods of Hirst [6] and tie them up with later more algebraic approaches. In particular we shall give a very satisfactory geometric explanation for the fundamental observation of Chasles that it is straightforward to compute the number of exceptional elements in a system satisfying given geometric conditions and that the knowledge of these elements, via degeneration formulas, can be used to find all elements satisfying the same geometric conditions. It is the hope that the beautiful geometric insights of the early years of the subject will not only clarify the enumerative formulas, but will also be of help in todays study in this field and that the general aspects of this field will motivate further investigations. Indeed, the point of view of "completing" geometric objects by considering not only the object itself and its deformations, but simultaneously to consider the associated osculating spaces and their deformations, is a valuable addition to geometry and deserve attention.

We have for simplicity of notation chosen to treat only the case of planes. The generalizations to higher dimensional spaces do not, in our approach, add serious difficulties and we shall return to the general treatment in the article with Lascoux mentioned above.

As indicated above the theory of complete correlations is parallelled by the theory of complete quadrics. These theories can be given virtually identical treatments. We refer to the articles of Battaglini [1], Chasles [2], [3], Severi [27], [28], Semple [24], [26], Study [29], Tyrrell [33] and Van der Waerden [38] for treatments similar to the presentation of this article and to Demazure and Vainsencher [34] for a more recent "functorial" approach.

2. The idea of complete correspondences is an abstraction of the correspondences that can appear when two planes are put into perspective with each other in all possible ways, and the limiting positions are studied by letting the center of projection approach one of the two planes or the line of intersection.

Given two planes S_2 and S_2' in a 3-dimensional space and let $\ell = \ell'$ be the line of intersection.

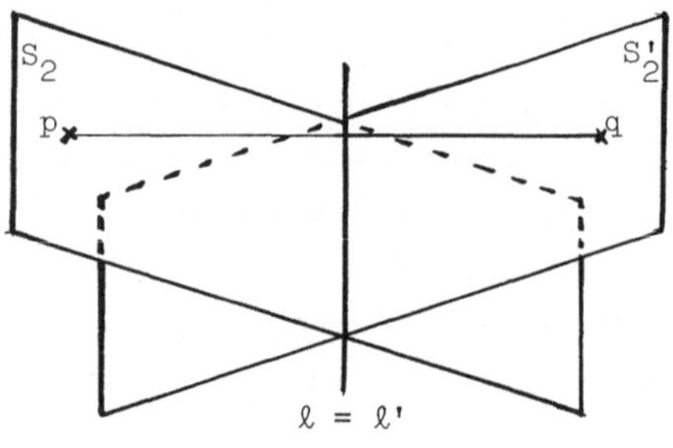

If we choose the center of perspective outside of the two planes we get a non-degenerate linear map, called a collineation, between the planes. A choice of perspective in the planes gives two different cases according to the center being on or outside the line $\ell = \ell'$. In each case we shall in the traditional way describe the correspondences by their effect on points and lines in the two planes.

I. The center of perspective p lies in S_2, but not in S_2'. Then we see that the perspective of a point q' of S_2' which is situated on ℓ' is the whole line pq'. This was in the classical literature expressed by saying that to the point q' corresponds an indeterminate point in S_2 on the determinate line pq'. The perspective of every other point in S_2' is the point p itself.

Similarly the perspective of the point p in S_2 is any point in S_2'. This is expressed by saying that the point p corresponds to a wholly indeterminate point of S_2'. To any other point q of S_2 corresponds the intersection between ℓ' and the line pq.

With respect to correspondences between lines we see that to the line ℓ' in S_2' corresponds a wholly indeterminate line in S_2 and to every other line m' in S_2' corresponds the line through p and the point of intersection $\ell \cdot m'$. Similarly, to every line m in S_2 which passes through the point p corresponds an indeterminate line through the point of intersection $\ell' \cdot m$ and to every other line in S_2 corresponds the line ℓ' itself.

In terms of projections the above correspondences can be described in the following way:

First, we are given a projection $\pi'\colon S_2' \dashrightarrow S_2$ with center on ℓ' and image p together with a collineation $\gamma'\colon \ell' \longrightarrow S_2/p$ where S_2/p is the target of the canonical projection $\lambda'\colon S_2 \dashrightarrow S_2/p$ with center p.

To a point q' on ℓ' corresponds under this correspondence an indeterminate point on the determinate line $\lambda^{-1}\gamma'(q')$ through p and to any other point corresponds the point $\pi'(q') = p$.

Secondly, we are given a projection $\pi\colon S_2 \dashrightarrow S_2'$ with center p and image ℓ' together with a collineation $\gamma\colon p \longrightarrow S_2'/\ell'$ where S_2'/ℓ' is the target of the canonical projection $\lambda'\colon S_2' \dashrightarrow S_2'/\ell'$ with center ℓ'.

To the point p corresponds, under such a correspondence, a wholly indeterminate point on the plane $(\lambda')^{-1}\gamma(p) = S_2'$ and to any other point q of S_2 corresponds the point $\pi(q)$.

II. <u>The center of perspective</u> $p = p'$ <u>is on the line</u> $\ell = \ell'$ <u>of intersection</u>. Then we see that to the point $p = p'$ regarded as a point in either plane corresponds a wholly indeterminate point in the other plane and to any other point on the line $\ell = \ell'$ in either plane corresponds an indeterminate point on the same line in the other plane. To any other point in either plane corresponds the point $p = p'$ itself.

As for lines, the line $\ell = \ell'$ considered as a line in either plane corresponds a wholly indeterminate line in the other plane and to any other line through the point $p = p'$ in either plane corresponds an indeterminate line through $p = p'$ considered as a point in the other plane. To any other line in either plane corresponds the line $\ell = \ell'$ considered as a line in the other plane.

In terms of projections the above correspondence can be described as a projection $\pi: S_2 \dashrightarrow S_2'$ with center ℓ and image p' together with a projection $\gamma: \ell \dashrightarrow S_2'/p'$ with center p and image the point corresponding to ℓ' by the canonical projection $\varepsilon': S_2' \dashrightarrow S_2'/p'$, and a collineation $\delta: p \xrightarrow{\sim} S_2'/\ell'$.

To the point p corresponds, under such a correspondence, a wholly indeterminate point in the inverse image S_2' of $\delta(p)$ by the canonical projection $S_2' \dashrightarrow S_2'/\ell'$ with center ℓ'. To any other point q on ℓ corresponds an indeterminate point on the determinate line $(\varepsilon')^{-1}\gamma(q) = \ell'$. Finally, to any point q in S_2 not on ℓ corresponds the point $p' = \pi(q)$.

To obtain a complete picture of the above description of collineations in terms of projections we should also have described the effect of the correspondences on lines. For the sake of brevity we leave this as a pleasant excercise for the reader.

To pass from correspondences defined by perspectives to general <u>homographic correspondences</u> or <u>complete collineations</u>, we merely have, following Hirst, to consider the planes as separate spaces not necessarily imbedded in any space. We then see that, corresponding to the two cases above, the <u>exceptional or degenerate collineations</u> can be described as follows:

I. There is a <u>singular point</u> in one plane and a <u>singular line</u> in the other whose correspondents are wholly indeterminate. To each point in the singular line corresponds an indeterminate point in a determinate line passing through the singular point. Conversely, to every line through the singular point corresponds an indeterminate line through a determinate point in the singular line.

II. In each plane there is a singular line and a singular point on that line. The correspondent of a singular point is wholly indeterminate and the correspondent of every other point on a singular line in one plane is an indeterminate point in the other singular line. Similarly, the correspondent of a singular line is wholly indeterminate and the correspondent of any other line through the singular point is an indeterminate line through the other singular point.

The description of complete correlations in terms of linear maps between not necessarily embedded planes similar to the above description in terms of projections is clear.

To pass from <u>collineations</u> to <u>correlations</u> we compose a collineation between two planes S_2 and S_2'' with a fixed non-degenerate correlation between S_2'' and S_2'. In order to translate the description of collineations to a description of correlations it is convenient to use the traditional terminology that a line in one plane corresponding to a point in the other under a correlation is called a <u>polar</u> of the point and a point corresponding to a given line is called a <u>pole</u> of the point. We obtain three different types of exceptional correlations as case I above present two varieties.

I. The exceptional correlation between S_2 and S_2' is such that in each plane there is a singular point whose polars are wholly indeterminate. The pole of every line through a singular point is an indeterminate point on a determinate line through the other singular point.

In terms of linear maps such a correlation is described as a linear map $\pi: S_2 \dashrightarrow S_2'^*$ with center the singular point p of S_2 and image the pole p'^* of the singular point p' in S_2', together with a non-degenerate correlation $\gamma: p \rightarrow S_2'^*/p'^*$.

I*. The exceptional correlation between S_2 and S_2' is such that in each plane there is a singular line whose poles are wholly indeterminate. The polar of every point in a singular line is an indeterminate line through a determinate point of the other singular line.

In terms of linear maps such a correlation is described as a linear map $\pi: S_2 \dashrightarrow S_2'^*$ with center on the singular line ℓ of S_2 and image the pole ℓ'^* of the singular line ℓ' of S_2', together with a non-degenerate correlation $\gamma: \ell \rightarrow S_2'^*/\ell'^*$.

II. The exceptional correlation between S_2 and S_2' is such that in each plane there is a singular line and a singular point on that line such that the polar of each singular point as well as the pole of each singular line is wholly indeterminate. The polar of any point on the singular line different from the singular point is an indeterminate line through the other singular point and the pole of any line different from the singular line is an indeterminate point in the other singular line.

In terms of linear maps such a correlation is described by a linear map $\pi: S_2 \dashrightarrow S_2'^*$ with center on the singular line ℓ of S_2 and image the pole ℓ'^* of the singular line ℓ' of S_2, together with a linear map $\gamma: \ell \dashrightarrow S_2'^*/\ell'^*$ with center on the singular point p of S_2 and image the point on $S_2'^*/\ell'^*$ corresponding to the pole p'^* of the singular point p' of S_2' and a correlation $\delta: p \longrightarrow S_2'^*/p'^*$.

We shall refer to these three types of exceptional correlations as exceptional correlations with singular points, singular lines and singular points and lines. We see that in terms of linear maps a complete correlation between two spaces S_d and S_d' can be defined as two sequences of linear spaces $L_r \subsetneqq L_{r-1} \subsetneqq \cdots \subsetneqq L_1 \subsetneqq L_0 = S_d$ and $L_r' \subsetneqq L_{r-1}' \subsetneqq \cdots \subsetneqq L_1' \subsetneqq L_0' = S_d'$ with $\dim L_i = \dim L_i'$ together with linear maps

$$L_i \dashrightarrow S_d'^*/L_i'^*$$

with center L_{i+1} and image $L_{i+1}'^*/L_i'^*$ for $i = 0,1,\ldots,r-1$, where $L_i'^*$ is the $(d-\dim L_i-1)$ dimensional linear space in $S_d'^*$ corresponding to the space L_i' of S_d'.

3. The description of degenerate correlations of the previous section together with the picture of deformation of correlations by letting the centre of perspective approach the two planes in 3-dimensional space was a sufficiently flexible tool in the hands of the early geometers to handle the degenerate correlations as codimension one subfamilies of one single family of complete correlations and to study their geometry and their enumerative properties. For example to determine the dimensions of the families of correlations between S_2 and S_2' with singular lines, it is sufficient to note that these correlations are determined when the singular points are determined together with three pairs of conjugate lines passing through them (in terms of linear maps a projection $S_2 \dashrightarrow S_2'^*$ with center p and image p'^* is determined by the points p and p' and a non-degenerate correlation $S_2/p \longrightarrow p'^*$ of lines, the latter being determined by the three pairs of conjugate

points corresponding to the three pairs of lines). Hence these correlations form a 2 + 2 + 3 = 7 dimensional family. Similarly, the correlations with singular lines are determined by the singular lines and three pairs of conjugate points on them, again a family of dimension 7. On the other hand the correlations with singular points and lines are completely determined by the location of the singular lines and the singular points on them and therefore form a family of dimension 2 + 2 + 1 + 1 = 6.

To give another example of how the geometric description was used to treat families of correlations we consider the following four conditions on correlations:

<u>A given point shall have a given polar</u>.

<u>A given line shall have a given pole</u>.

(E) <u>Two given points shall be conjugate</u>.

<u>Two given lines shall be conjugate</u>.

Of these the two last impose one condition on the correlations, that is the families μ, respectively ν, that satisfy the third, respectively the fourth condition form closed codimension one families in the space of complete correlations. The two first consequently impose two conditions each.

It is to be noticed that the conditions are to be considered as restrictions both on the non-degenerate and on the exceptional correlations. The meaning of the conditions on non-degenerate correlations is clear. To see what they mean for exceptional ones we consider as an example the condition that the points q in S_2 and q' in S_2' be conjugate under a correlation with singular points p and p'. Then either p = q or, if $p \neq q$, we must have that the pole of q which is a determinate line through p' passes through q'. For the points q and q' to be conjugate under a correlation with singular lines ℓ and ℓ' we must have that either q is in ℓ or, if q is not in ℓ, then ℓ' must pass through q'.

Geometric conditions of the above type are called <u>elementary</u>. A one-dimensional family of correlations satisfying any seven independent geometric conditions is called a <u>system</u> of correlations and a family satisfying any seven independent elementary conditions is called a <u>fundamental system</u>.

Fix two points q in S_2 and q' in S_2' and a system C of correlations. The transforms of q by the correlations of the system is a curve C'^* of

$S_2'^*$. The polars of the point q under the correlations of the system are tangents of the dual curve C' of C'*. The class μ_C of the system C is defined to be the class of the curve C', that is the number of polars that pass through the point q'. Since each of the correlations under which the polar of q passes through q', the polar of q' passes through q, we see that the number μ_C is also the number of polars of q' that pass through q. Hence μ_C is the number of correlations of the system C in each of which q and q' are conjugate.

Similarly the poles of a fixed line ℓ in S_2 under the correlations of the system form a curve in S_2' whose dual C'* has as tangents all the lines in $S_2'^*$ corresponding to the line ℓ. The degree ν_C of the system C is defined to be the degree of the curve C'*, that is the number of poles that lie on a fixed line ℓ' of S_2'. Again for each correlation under which the pole of ℓ lies on ℓ', the pole of ℓ' lies on ℓ, so that ν_C is also the number of poles of ℓ' that lie on ℓ. Hence ν_C is the number of correlations of the system in each of which ℓ and ℓ' are conjugate.

Given a system C of correlations, the numbers μ_C and ν_C are called the characteristics of the system. They are clearly the same as the intersection numbers $\mu.C$ and $\nu.C$ where μ respectively ν are the correlations satisfying conditions (E) 3 respectively 4. If we denote by π_C respectively ν_C the number of exceptional correlations of the system C with singular points respectively singular lines, these numbers are related to the characteristics of C by the following relations:

$$\text{(D)}\quad \begin{aligned} \mu_C &= 2\nu_C - \pi_C \\ \nu_C &= 2\mu_C - \lambda_C \end{aligned}$$

We note that the system C of correlations will, in general not contain an exceptional correlation with singular points and lines, because these form a family of codimension two.

To prove the formulas (D) Hirst [6] uses the coincidence formula. First we note that the relations are dual so that it suffices to prove the first relation. Fix two lines ℓ_1 and ℓ_2 in S_2 and a point q' in S_2' and let q denote the intersection point $\ell_1\ell_2$. By hypothesis there are ν_C correlations of the system in each of which the pole p_1' of ℓ_1 lies in an arbitrary line m' through q'. If the pole p_2' of ℓ_2 under each of these ν_C correlations is joined to the point q' by a line n', then to each line m' in the pencil of lines through q' there will correspond ν lines n' in the same pencil. Similarly, to each line n' through q' there will correspond ν lines m' in the pencil of lines through q' that contain the poles of ℓ_1 under the ν correlations in which the poles of ℓ_2 lie in n'. This establishes a (ν,ν)-relation on the pencil of lines through q' and by the coincidence formula there are exactly 2ν lines in each of which an m' and its corresponding n' coincide.

Each of the 2ν lines passes through the poles of ℓ_1 and ℓ_2 in the same correlation. If the correlation is non-degenerate, then this line is the polar of the intersection q because the polar passes through the poles of ℓ_1 and ℓ_2. Also if the correlation is degenerate with singular lines ℓ and ℓ' then the poles of the lines ℓ_1 and ℓ_2 are different points in the line ℓ' and the polar of q is the line ℓ'. However, if the correlation has singular points it is not in general true that each of the 2ν lines is the polar of q in some correlation although they contain the poles of ℓ_1 and ℓ_2. Indeed, if p and p' are the singular points of one of the π_C exceptional correlations with singular points included in the system C, then the poles of the lines ℓ_1 and ℓ_2 will coincide with p', because as we have seen the pole of every line in S_2, that do not pass through the singular point, coincides with the singular point in S_2. Hence the line m' = n' coincides with the line q'p'. The polar of the intersection point q in this exceptional correlation will not, however, in general coincide with q'p', but with the line through p' which corresponds to the line joining q and p in the correspondence between the pencil of lines through p and the pencil of lines through p' in which a line m corresponds to the determinate line through p' in which the poles of m are situated. Indeed, the polar of q is a determinate line passing through the singular point p' and therefore must coincide with the line qp.

We have proved that among the 2ν correlations under which the lines joining the poles of ℓ_1 and ℓ_2 with q' coincide, that is those for which the polar of q can pass through q', there are exactly the π_C exceptional correlations with singular points in which the polar of q does not pass through q'. Hence we have proved that $2\nu_C - \pi_C$ is equal to the number μ_C of correlations in which the polar of q passes through q'.

The goal of Hirsts [6] investigations was to determine the number of correlations satisfying any eight independent conditions. To accomplish this he uses the ingenious observation, already made by Chasles in the similar case of conics, that the number of exceptional elements in a system can often be determined directly by geometric arguments and then the caracteristics of the system can be determined by the degeneration relations (D). By this method he determines the number of correlations satisfying any eight elementary conditions by first determining the exceptional correlations in any fundamental system. The fundamental systems correspond to all solutions of the equation

$$2a + 2b + c + d = 7$$

in natural numbers. There are 52 such solutions. The fourtuple (a,b,c,d) determines the fundamental system and is called the signature of the system. Since (a,b,c,d) and (b,a,c,d) are indentical and (a,b,c,d) and (b,a,d,c) are dual there are only 13 systems to be considered. Hirst determines the exceptional correlations in all these systems by beautiful geometric arguments. We shall indicate his arguments in the cases (0,0,d,7-d) only. The reason for restricting our attention to these cases is that it is clear that the four elementary conditions (E) are equivalent to the conditions μ^2, ν^2, μ, ν so that all the fundamental systems have the same number of exceptional correlations as the systems $\mu^d\nu^{7-d}$ for d = 0,1,2,3 and their duals. It is therefore remarkable that Hirst really was interested in the finer structure given by the signature. The more so as the conditions ν and μ form a basis for all conditions on correlations. Schubert [20] realized the importance of the systems $\mu^d\nu^c$ and saw how to make Hirsts method for determining the systems (0,0,c,d) into an inductive scheme for generalizing the results to higher dimensional spaces.

To illustrate Hirsts and Schuberts [20] methods we first consider the case ν^7. The number $\pi\nu^7$ is zero because if we choose seven lines in general position in each plane that shall be pairwise conjugates under a correlation with singular points, then the singular point in one plane can be in at most two lines. The pole of any line not passing by the singular point is the singular point in the other plane. Hence the poles of at least five lines is the singular point in the other plane, contradicting the generality of the conjugate lines.

The number $\lambda\nu^7$ of correlations in the system ν^7 with singular lines is equal to the number $(1,2)\nu^7$ of correlations between two lines in two planes under which the intersection points of the line with seven given lines in one plane are conjugate with the intersection points of the other line with seven given points in the other plane. Indeed, the seven pairs of lines that define the system ν^7 are conjugate under a correlation with singular lines if and only if their poles, that lie on the singular line in the other plane are conjugats under the correlation induced on the singular lines. The number $(1,2)\nu^7$ is easily seen to be 3. This is also a consequence of Sturms work on groups of points on the line.

Now the number ν^8 can be computed by the degeneration relation $3\nu_C = 2\pi_C + \lambda_C$ which is a consequence of the relations (D). We obtain $3\nu^8 = 2\pi\,\nu^7 + \lambda\,\nu^7 = 3$ so that $\nu^8 = 1$. Thus we have that $(0,b,0,8-2b) = (a,0,8-2a,0) = 1$. More generally it is clear from Hirsts computations of the systems with signature $(0,0,c,d)$ and $c + d = 7$ that in order to compute the numbers $\lambda\mu^c\,\nu^d$ one can restrict to the case of points on the line and compute the number $(1,2)\mu^c\,\nu^d$ of correlations of two plane lines such that d pairs of points are conjugate and which satisfy c times the condition μ restricted to the lines. The latter condition is equivalent to twice the simple Schubert condition that a line passes through a fixed point. We shall not go into details here as this point will be made precise in section 5.

Schubert observed that in view of the above computations and the degeneration formula $\mu_C = 2\mu_C - \lambda_C$ it is possible to compute the numbers $\mu^c\,\nu^d$ for $c + d = 8$ by descending induction on d starting with $d = 8$ and the case of points on the line. Indeed, one obtains that $2\,\mu^c\,\nu^d = \mu^{c-1}\nu^{d+1} + \lambda\mu^{c-1}\nu^d = \mu^{c-1}\nu^{d+1} + (1,2)\mu^{c-1}\nu^d$. Knowing that $(1,2)\mu\nu^6 = 6$, $(1,2)\mu^2\nu^5 = (1,2)\mu^3\nu^4 = 12$ one obtains $\mu\nu^7 = 2$, $\mu^2\nu^6 = 4$, $\mu^3\nu^5 = 8$ and $\mu^4\nu^4 = 10$, that (up to a printing error in [1,0,3,3] in Hirst [6], p.62) are the numbers given by Hirst.

With the above background it is easy to understand Schuberts generalization of the previous results to higher dimensions. First of all he could use the generalization of the concept of complete correlations to higher dimensions that seems to be part of the folklore at the time. Secondly, he needed degeneration formulas of the above type (D) for correlations in higher dimensional spaces in a common ambient spaces. Such formulas follow from coincidence formulas, or correspondence principles, by methods similar to the one used by Hirst for proving the formulas (D), and that we indicated above. The methods for obtaining such formulas from correspondence principles were indeed standard at the time and used e.g. by Chasles [2] in his theory of conics. The generalized correspondence principle that Schubert needed in this case also seems to be part of the folklore of the time and follows, in Schuberts own words, easily from the original one for points on lines (or dually as we used them for lines through a point). The degeneration formulas that Schubert obtained are completely similar to those obtained by Hirst, but the one corresponding to conjugate points contains a term corresponding to twice a simple Schubert condition, like the term μ in the expression $(1,2)\mu^c\nu^d$ above. A third part of the generalization was the introduction of Schubert conditions. These impose themselves quite naturally, as we have seen, and do rather simplify the induction procedure than impose additional difficulties. To start the induction Schubert also had to treat the most degenerate case that is of flags of linear spaces satisfying characteristic conditions similar to the conditions μ and ν. The main contribution of Schubert is, however, focusing on the conditions given by the characteristics of the system and the setup of the inductive procedure together with finding the functions that solve the problem in particular cases. We shall not go into the procedure as it follows exactly along the lines explained above and for the actual functions found by Schubert we refer to the forthcoming article with Lascoux. Actually Schubert did not perform the calculations in the case of correlations. He merely announced the formulas and performed the similar calculations for quadrics. The proofs were supplied by Giambelli [4] along the lines indicated by Schubert, however, Giambelli was able to improve the combinatorial arguments considerably and to generalize Schuberts formulas.

4. The above examples illustrate the ease and elegance with which exceptional correlations were treated during the early days of the subject and how they considered systems of correlations within the space of complete correlations. They had an excellent understanding of the limit positions of families of non-degenerate correlations when they approach the exceptional correlations (see also Del Prete [5], Predella [13], Segre [22], [23]). To later generations it has, however, appeared necessary to construct explicit parameter varieties for the

complete correlations and to study the local structure of such varieties. A wide variety of models for the parameter space have been constructed. We shall expose the method used by Tyrrell [33] and scetched by Semple [25] and which gives the most complete results in a simple manner.

We shall denote the (projective) space of all linear maps from S_2 to $S_2'^*$ by $L = L(S_2, S_2')$. The subvarietees of linear maps of rank 2 respectively rank 1 we denote by M_1 respectively M_2. Clearly, M_1 is of codimension 1 and is non-singular outside of M_2, and M_2 is a non-singular subvariety of dimension 4. The natural way of studying the limiting positions of systems of non-degenerate linear maps when they approach the loci M_1 and M_2 of degenerate linear maps is by separating the normal directions. Since there is just one normal direction to each point of $M_1 - M_2$ it is sufficient to separate the normal directions to M_2 in L. This is done by constructing the monoidal transformation of L with center on M_2. To this end we let $L_1 = L_1(S_1, S_1')$ be the space of second adjugates of the linear maps from S_1 to $S_1'^*$, that is, if we choose bases for S_1 and S_1' the space L_1 is the space of matrices whose coordinates are the 2×2-minors of the matrices representating the points of L with respect to the given bases. We shall throughout choose a basis for the space L_1 by ordering the rows and columns of its points lexicographically.

The natural map

$$L - M_1 \longrightarrow L_1$$

which sends the 3×3-matrix α to a 3×3-matrix whose $((i,j), (i',j'))$'th coordinate is the 2×2-minor of A taken from i and j and columns i' and j', defines an immersion

$$L - M_1 \longrightarrow L \times L_1 .$$

The closure Ω of the image of this immersion is (a twisting of) the monoidal transform of L with center on M_2.

We shall show that Ω is a parameter space for the complete correlations in such a way that the exceptional locus represents the correlations with singular lines, the closure of the image of $M_2 - M_1$ represents the correlations with singular points and the correlations with singular points and lines are the intersection of the other two exceptional loci.

Our study of the space Ω shall be local. To find a convenient covering of the space Ω we note that the coordinate transforms of S_2 and $S_2'^*$ operate on the spaces L and L_1, and it is clear that for a general choise of coordinate transforms we can, given a point (α,α_1) in $L \times L_1$, make any coordinate of the matrices α and α_1 non-zero. Hence we may assume that (α,α_1) lies in the open affine piece $V \times V_1$ of matrices whose coordinate in the upper left corner is 1. We fix such a coordinate system.

Let S respectively T be the affine spaces of upper respectively lower diagonal matrices with all diagonal entries equal to one and let U be the space of diagonal matrices with diagonal elements of the form $(1,c_1,c_1c_2)$. We shall write such matrices as $\mathrm{diag}(1,c_1,c_1c_2)$. The coordinate functions of the spaces S and T we denote by $s_{i,j}$ and $t_{i,j}$ and we consider U as the open subset of the affine space with coordinate functions u_1 and u_1u_2, where the function u_2 is non-zero.

The natural map

$$T \times U \times S \longrightarrow L$$

which sends the matrices (B,C,A) of $T' \times U \times S$ to their product BCA is given on coordinate functions by sending the coordinate matrix $(x_{i,j})$ of L (where $x_{1,1} = 1$) to the matrix

$$(*) \quad \begin{bmatrix} 1 & s_{0,1} & s_{0,2} \\ t_{1,0} & t_{1,0}s_{0,1}+u_1 & t_{1,0}s_{0,2}+u_1s_{1,2} \\ t_{2,0} & t_{2,0}s_{0,1}+u_1t_{2,1} & t_{2,0}s_{0,2}+t_{2,1}s_{1,2}u_1+u_1u_2 \end{bmatrix}$$

We note that the determinant $\Delta_2 = x_{1,1} - x_{1,0}x_{0,1}$ is mapped to u_1 and the determinant Δ_3 of $(x_{i,j})$ is mapped to $u_1^2u_2$ so that we obtain an isomorphism

$$k[x_{i,j},\Delta_2^{-1},\Delta_3^{-1}] \longrightarrow k[s_{i,j},t_{j,i},u_1,u_2,u_1^{-1},u_2^{-1}].$$

or what is the same an isomorphism

$$(x) \quad T \times U_{u_1,u_2} \times S \longrightarrow L_{\Delta_2,\Delta_3}$$

where the subscripts means the open sets where the functions are non-zero.

Making use of the formula $\overset{2}{\Lambda}(t_{j,i})(u_k)(s_{i,j}) =$ $= \overset{2}{\Lambda}(t_{j,i})\cdot\overset{2}{\Lambda}(u_k)\cdot\overset{2}{\Lambda}(s_{i,j})$ we obtain that the matrix $\overset{2}{\Lambda}(x_{i,j})$ is mapped to u_1 times the matrix

$$(*)_1 \quad \begin{bmatrix} 1 & s_{1,2} & s_{0,1}s_{1,2}-s_{0,2} \\ t_{2,1} & t_{2,1}s_{1,2}+u_2 & t_{2,1}s_{0,1}s_{1,2}-t_{2,1}s_{0,2}+s_{0,1}u_2 \\ t_{1,0}t_{2,1}-t_{2,0} & t_{1,0}t_{2,1}s_{1,2}-t_{2,0}s_{1,2}+t_{1,0}u_2 & v \end{bmatrix}$$

where $v = (t_{1,0}t_{2,1}-t_{2,0})(s_{0,1}s_{1,2}-s_{0,2}) + t_{1,0}s_{0,1}u_2 + u_1u_2$.
Sending the coordinates $(y_{i,j})$ of L_1 onto the corresponding coordinates of the matrix $(*)_1$ we obtain a map

$$T \times U \times S \longrightarrow V_1$$

The product map

$$(**) \quad T \times U \times S \longrightarrow V \times V_1$$

is seen from the expressions $(*)$ and $(*)_1$ to be a closed immersion with image defined by the equations

$$(F) \quad \begin{aligned}
x_{1,2} &= x_{1,0}x_{0,2} + \Delta_2 y_{0,1} \\
x_{2,1} &= x_{2,0}x_{0,1} + \Delta_2 y_{1,0} \\
x_{2,2} &= x_{2,0}x_{0,2} + \Delta_2 y_{1,0}y_{0,1} + \Delta_2 \cdot \Delta_2^1 \\
y_{2,0} &= x_{1,0}y_{1,0} - x_{2,0} \\
y_{0,2} &= x_{0,1}y_{0,1} - x_{0,2} \\
y_{2,1} &= x_{1,0}y_{1,0}y_{0,1} - x_{2,0}y_{0,1} + x_{1,0}\Delta_2^1 = x_{1,0}y_{1,1} - x_{2,0}y_{0,1} \\
y_{1,2} &= x_{0,1}y_{1,0}y_{0,1} - x_{0,2}y_{1,0} + x_{0,1}\Delta_2^1 = x_{0,1}y_{1,1} - x_{0,2}y_{1,0} \\
y_{2,2} &= (x_{1,0}y_{1,0}-x_{2,0})(x_{0,1}y_{0,1}-x_{0,2}) + x_{1,0}x_{0,1}\Delta_2^1 + \Delta_2\,\Delta_2^1
\end{aligned}$$

where $\Delta_2^1 = y_{1,1} - y_{1,0}y_{0,1}$.

From the previous remarks it is clear that the above map $L - M_1 \longrightarrow L \times L_1$ restricts to the map obtained by composing the inverse of the map (x) with the restriction of the map $(**)$. Consequently, the above equations define $\Omega \cap (V \times V_1)$.

Choose a point $\alpha \times \alpha_1$ in $\Omega \cap (V \times V_1)$. Then we can write $\alpha = BCA$ where (B,C,A) is a point in $T \times U \times S$. We see from the matrices $(*)$ and $(*)_1$ above that by changing the coordinates of S_2 respectively $S_2'^*$ by the inverse of the matrix B respectively A, the point $\alpha \times \alpha_1$ takes the form $\mathrm{diag}(1,c_1,c_1c_2) \times \mathrm{diag}(1,c_2,c_1c_2)$. By a further non-singular "diagonal" correlation we see that the complete correlations are projectively equivalent to one of the following four types:

$$(N) \quad \begin{array}{ll} \mathrm{diag}(1,1,1) \times \mathrm{diag}(1,1,1) & \text{corresponding to } c_1 \neq 0 \neq c_2, \\ \mathrm{diag}(1,1,0) \times \mathrm{diag}(1,0,0) & \text{corresponding to } c_1 \neq 0 = c_2, \\ \mathrm{diag}(1,0,0) \times \mathrm{diag}(1,1,0) & \text{corresponding to } c_1 = 0,\ c_2 \neq 0, \\ \mathrm{diag}(1,0,0) \times \mathrm{diag}(1,0,0) & \text{corresponding to } c_1 = 0 = c_2. \end{array}$$

We see that the first case corresponds to non-degenerate correlations. The second case corresponds to the correlations with singular points. As far the third case, the map $\mathrm{diag}(1,0,0)$ gives a projection with center on a line and image a point and the map $\mathrm{diag}(1,1,0)$ then corresponds to a non-degenerate correlation from the line to the target of the projection with center on the point. Hence, the third case corresponds to correlations with singular lines. Similarly, the fourth case corresponds to correlations with singular points and lines.

We have proved that the space Ω parametrizes the complete correlations between two planes and that there are exactly 4 projectively distinct types of complete correlations whose normal forms are given in (N) above.

From the above results we have that the subvarieties π respectively λ of Ω defined as the closures of the hyperplanes $(q_2 = 0) \cap \Omega$ respectively $(q_1 = 0) \cap \Omega$ parametrize the exceptional correlations with singular lines respectively the exceptional correlations with singular points. They are called the degeneration mainfolds of the first kind and their intersection the degeneration mainfold of the second kind. The definition of π and λ given above depends on the coise of coordinates in the affine space $T \times U \times S$ which is not invariant. However, it is easy to show that the degeneration manifolds are invariantly defined. We omit the proof since we shall give a coordinate free approach in the next section. The main result about the geometry of Ω is:

<u>The parameter space Ω of complete correlations is non-singular of dimension 8. Its primary degeneration manifolds π and λ are non-singular divisors intersecting transversally in the variety of second order degenerations</u>.

We have proved everything except the non-singularity and the transversality. Since the intersecetion of π and λ are defined by the equations $q_1 = q_2 = 0$ it is sufficient to prove that the jacobian of the system q_1, q_2 does not drop rank at any point. However, from the equations (F) for the variety $\Omega \cap (V \times V_1)$ it is clear that the tangent space of this variety has a basis $\frac{\partial}{\partial x_{0,1}}, \frac{\partial}{\partial x_{1,0}}, \frac{\partial}{\partial x_{0,2}}, \frac{\partial}{\partial x_{2,0}}, \frac{\partial}{\partial x_{1,1}}$, $\frac{\partial}{\partial y_{0,1}}, \frac{\partial}{\partial y_{1,0}}, \frac{\partial}{\partial y_{1,1}}$, at every point and that the subspace

$$\left\{\frac{\partial}{\partial x_{1,1}} - x_{1,0}\frac{\partial}{\partial x_{0,1}} - x_{0,1}\frac{\partial}{\partial x_{1,0}}, \frac{\partial}{\partial y_{1,1}} - y_{1,0}\frac{\partial}{\partial y_{0,1}} - y_{0,1}\frac{\partial}{\partial y_{1,0}}\right\}$$

corresponding to the equations $q_1 = (x_{1,1} - x_{0,1}x_{1,0})$ and $q_2 = (y_{1,1} - y_{0,1}y_{2,0})$ is two-dimensional at every point.

The results of this section provide a complete algebraic description of the complete correlations and makes it possible to redo the results of the previous sections in a more algebraic manner. We shall illustrate this by giving a proof of the degeneration relations (D) for divisors. To this end we first describe the cycles μ respectively ν of correlations under which two given points, respectively two given lines are conjugate. Fix a point q in S_2 and a point q' in S_2'. Then μ is defined by the subvariety of Ω of complete correlations $\alpha \times \alpha_1$ such that the composite linear map

$$q \longrightarrow S_2' \xrightarrow{\alpha} S_2'^* \dashrightarrow S_2'^*/q'^*$$

is not defined. In other words μ can be defined by the vanishing of any fixed coordinate of the maps in L.

Similarly, fix a line m in S_2 and a line m' in S_2'. The lines m and $S_2'^*/m'^*$ define a coordinate in the space L_1 of second adjugates. Then ν is defined by the subvariety of Ω of complete correlations (α, α_1) such that the corresponding coordinate of α_1 is zero. If (α, α_1) corresponds to a non-degenerate correlation or a correlation with singular points then α_1 is the second adjugate of α so that (α, α_1) is in this variety if and only if the composite linear map

$$\ell \to S_2 \xrightarrow{\alpha} S_2'^* \longrightarrow S_2'^*/\ell'^*$$

is degenerate.

To see that the above definitions of μ and ν impose the same geometric conditions on the exceptional correlations as those given in the previous section we note that for a correlation (α,α_1) with singular points p and p' we have that p is the point where α is not defined and p'^* is the image of α. Hence (α,α_1) is in μ if $p = q$ or if $p \neq q$ and $\alpha(q)$ which is in the line p'^* also lies on q'^*. For a correlation (α,α_1) with singular lines ℓ and ℓ' we have that ℓ is the kernel of α and ℓ'^* is the image of α. Hence (α,α_1) is in μ if and only if q is in ℓ, or if q is not in ℓ then $\alpha(q)$, which is the point ℓ'^*, is in q'^*, that is q' is in ℓ'. The conditions defining μ are consequently the same as the geometric ones of the previous section. Those for ν are treated similarly.

We have seen that the cycle μ and ν can be defined by the vanishing of any coordinate function appearing as a coordinate in the matrix $(*)$ and $(*)_1$ respectively. The matrix $(*)_1$ multiplied by a μ_1 is equal to the second adjugate of $(*)$ and since the coordinates of the second adjugate defines the cycle 2μ, and q_1 the cycle λ, we obtain that

$$2\mu = \nu + \lambda.$$

Similarly, one obtains the relation

$$2\nu = \mu + \pi.$$

It is not difficult to give proofs of the degeneration formulas in the setting of this section that are more similar to the one of the previous sections.

5. We shall in the following take a more global view of the construction of the preceding section which leads to a complete and very satisfactory description of the parameter space Ω. As a result we obtain a natural explanation of the original method of Chasles, as used by Hirst for correlation, which consists in first finding the exceptional correlations of a system satisfying certain elementary conditions and then using the degeneration relations to compute all

correlations satisfying the elementary conditions. In a different terminology it corresponds to computing the equivalence classes of a monoïdal transformation by restricting to the exceptional locus which in the case at hand is a projective bundle over the center of the transformation and therefore easy to handle. We also get a complete picture of Schuberts inductive procedure of determining the correlations satisfying elementary conditions.

Let E and E' be vector spaces of dimension 3 and let $S_2 = \mathbb{P}(E)$ and $S_2' = \mathbb{P}(E')$. As in the previous section we let $L = L(S_2, S_2'^{*})$ be the projective space of linear maps from E* to E' so that $L = \mathbb{P}(\mathrm{Hom}(E^*, E')^*) = \mathbb{P}(E^* \otimes E'^*)$. We then have that $L_1 = \mathbb{P}(\overset{2}{\Lambda} E^* \otimes \overset{2}{\Lambda} E'^*)$ is the space of second adjugates. The universal line bundles on L and L_1 we denote by N respectively N_1 and the universal maps by $u: E^* \longrightarrow E' \otimes N$ respectively $u_1: \overset{2}{\Lambda} E^* \longrightarrow \overset{2}{\Lambda} E'^* \otimes N_1$. The images I_2 and I_1 of the maps $v: \overset{2}{\Lambda} E^* \otimes \overset{2}{\Lambda} E'^* \otimes N^{-2} \to \mathcal{O}_L$ and $\overset{3}{\Lambda} E^* \otimes \overset{3}{\Lambda} E'^* \otimes N^{-3} \to \mathcal{O}_L$ resulting from the second and third exterior power of u define the subvarieties M_2 respectively M_1 of L. As above we denote by Ω the closure in $L \times L_1$ of the image of $L - M_1$ by the map $\mathrm{id}_L \times (v|L - M_1)$. Then the image of $I_2 \otimes \mathcal{O}_\Omega$ in $\mathcal{O}_\Omega$ is an invertible sheaf J_2 defining the subscheme $\Lambda = \Omega \times_L M_2 = \mathbb{P}(I_2/I_2^2 \otimes N^2)$ of Ω and $J_2 \otimes L$ is the restriction of N_1 to Ω, which we by abuse of language shall also denote by N_1.

Let

$$0 \to Q \to E \longrightarrow K \to 0$$
$$0 \to Q' \to E' \longrightarrow K' \to 0$$

be the universal sequences on the spaces $G = \mathrm{Grass}_1(E)$ and $G' = \mathrm{Grass}_1(E')$ of lines in $\mathbb{P}(E)$ respectively $\mathbb{P}(E')$. We see that $M_2 = G \times G'$, where a homomorphism $E^* \to E'$ is represented by the kernel K* and cokernel K' together with the (unique projective) linear map $Q^* \to Q'$. The embedding $M_2 \to L$ is clearly the Segre product. On M_2 the map $K^* \to E^* \to E' \otimes N$ is zero so that the image of $\overset{2}{\Lambda} K^* \otimes \overset{2}{\Lambda} E'^* \otimes N^{-2} \to \overset{2}{\Lambda} E^* \otimes \overset{2}{\Lambda} E'^* \otimes N^{-2} \to \mathcal{O}_L$ is in I_2^2. Similarly, the image of $\overset{2}{\Lambda} E^* \otimes \overset{2}{\Lambda} K'^* \otimes N^{-2} \to \mathcal{O}_L$ is in I_2^2. It follows that the map $\overset{2}{\Lambda} E^* \otimes \overset{2}{\Lambda} E'^* \to N^2 \otimes I_2$ factors via $\overset{2}{\Lambda} E^* \otimes \overset{2}{\Lambda} E'^* \to K^* \otimes Q^* \otimes K'^* \otimes Q'^*$ on M_2. However, I_2/I_2^2 is a locally free module on M_2 of rank 4 and is consequently isomorphic to $K^* \otimes Q^* \otimes K'^* \otimes Q'^*$. We picture the situation in the following Cartesian diagram

$$\begin{array}{ccc} \mathbb{P}(K^* \otimes Q^* \otimes K'^* \otimes Q'^*) & = \Lambda \longrightarrow & \Omega \\ \downarrow & & \downarrow \\ G \times G^1 & = M_2 \longrightarrow & L. \end{array}$$

Denote by $\bar{I}_2$ the ideal defining M_2 in M_1 and by Π the closure in Ω of the image of the map $(M_1 - M_2) \to L \times L_1$ defined by the surjection $(\overset{2}{\Lambda}E^* \otimes \overset{2}{\Lambda}E'^* \otimes N^{-2}) \longrightarrow \bar{I}_2$. As above the image of $\bar{I}_2 \otimes \mathcal{O}_\Pi$ in $\mathcal{O}_\Pi$ is an invertible ideal defining the subscheme $\Lambda \cap \Pi = \mathbb{P}(\bar{I}_2/\bar{I}_2^2)$ of Π.

It is easy to tie up the above construction with the one of the previous section. There we proved that Ω can be covered by affine pieces of the type $V \times V_1 \cap \Omega \cong T \times U \times S$ and that in the coordinate ring $k[T \times U \times S]$ of this piece the image of the map $\overset{2}{\Lambda}E^* \otimes \overset{2}{\Lambda}E'^* \otimes N^{-2} \to$ $\to I_2 \otimes \mathcal{O}_{T\times U\times S}$ is generated by the element u_1. Hence, the variety Λ of this section is the same variety as the closure of $V(u_1)$ defined in the previous section and represents the variety of degenerate correlations with singular lines. Moreover, we see that the image of $\overset{3}{\Lambda}E^* \otimes \overset{3}{\Lambda}E'^* \otimes N^{-3} \to \mathcal{O}_{T\times U\times S}$ is $(u_1^2 u_2)$ so that the image of $(M_1 - M_2) \cap V$ in $\Omega \cap (V \times V_1)$ is defined by $u_1^2 u_2$. Hence the closure of this image is defined by the kernel (u_2) of the map $k[T\times U\times S] \to (k[T\times U\times S]/u_1^2 u_2)_{u_1}$ so that the variety Π of this section is the same as the closure of $V(u_2)$ defined in the previous section and represents the correlations with singular points.

The implications to the enumerative aspects are now clear. By the definition of Chern classes we have that $[\Lambda] = \lambda = -c_1(J_2)$, $\mu = c_1(N)$ and $\nu = c_1(N_1)$. From the isomorphism $N_1 \cong J_2 \otimes N^2$ given above we obtain the degeneration relation

$$\nu = 2\mu - \lambda.$$

We see clearly the significance of the method used by Chasles and Hirst. To compute the class $\mu^i \nu^{8-i}$ they reduce to the easier task of computing the class $\lambda \mu^i \nu^{7-i}$ which is the restriction $\mu_\Lambda^i \nu_\Lambda^{7-i}$ of $\mu^i \nu^{7-i}$ to the projective bundle $\mathbb{P}(K^* \otimes Q^* \otimes K'^* \otimes Q'^*)$ over $G \times G'$ with relative hyperplane section ν_Λ and where $\mu_\Lambda = c_1(Q^*) + c_1(Q'^*)$. The latter computation is purely formal once the Chern classes of $K^* \otimes Q^* \otimes K'^* \otimes Q'^*$ are known and setting $\kappa = c_1(Q^*)$ and $\kappa' = c_1(Q'^*)$ we obtain $c(K^* \otimes Q^* \otimes K'^* \otimes Q'^*) = (1-\kappa-\kappa)(1-\kappa)^{-3}(1-\kappa')^{-3} =$ $= 1 + 2(\kappa+\kappa') + 3(\kappa^2+\kappa'\kappa+\kappa'^2) + 3\kappa\kappa'(\kappa+\kappa')$ which gives $\lambda\nu^7 = 3$, $\lambda\mu\nu^6 = 6$,

$\lambda\mu^2\nu^5 = \lambda\mu^3\nu^4 = 12$, $\lambda\mu^4\nu^3 = 6$ and $\lambda\mu^i = 0$ for $i > 4$. Also they compute the numbers $\pi\mu^i\nu^{7-i}$ that are easily seen to be zero for $i < 3$ because M_1 and Π are isomorphic up to codimension 3. It is sufficient to know that $\pi\nu^7 = 0$ because then ν^8 can be computed from the degeneration formula $3\nu = 2\pi + \lambda$ and then $\mu^i\nu^{8-i}$ can be computed inductively by repeated use of the formula $2\mu = \nu + \lambda$.

With the above picture the inductive procedure of Schubert also becomes clear. He observes that to compute $\lambda\mu^i\nu^{8-i}$ is the same as solving the problem of correlations for lines in 3-space where μ_Λ is interpreted as a sum of Schubert cycles. This observation is exactly what is expressed by the vector bundle $\mathbb{P}(K^* \otimes Q^* \otimes K'^* \otimes Q'^*) \to G \times G'$ where $G \times G'$ parametrizes pairs of lines (ℓ, ℓ') in $S_2 \times S_2'$ and where the fiber $\mathbb{P}(K^*(\ell) \otimes K'^*(\ell)) = \mathbb{P}(\mathrm{Hom}(K^*(\ell), K'(\ell))^*)$ parametrizes the correlations between the lines $\mathbb{P}(K(\ell))$ and $\mathbb{P}(K'(\ell))$. Here $\mu_\Lambda = c_1(Q^*) + c_1(Q'^*)$ is the sum of the Schubert cycles of lines through a given point and ν_Λ is the condition that a pair of points are conjugate under the correlation.

REFERENCES

[1] G. Battaglini: "Sui complessi ternari di primo ordine e di prima classe". Giorn. di Mat 20 (1882),230-248.

[2] M. Chasles: "Determination du nombre des sections coniques qui doivent toucher cinq courbes données d'ordre quelquonque, ou satisfaire à diverses autres conditions." C.R. de l'Acad. de Sciences 58 (1864), 222-226.

[3] M. Chasles: "Construction des coniques qui satisfont à cinq conditions. Nombre de solutions dans chaque question". C.R. 58 (1864), 297-308.

[4] G.Z. Giambelli: "Il problema della correlazione negli iperspazi". Mem. Reale Inst. Lombardo 19 (1903·), 155-194.

[5] G. Del Prete: "Le corrispondenze projettive degeneri". Rend. Inst. Lombardo 30 (1897), 400-409.

[6] T.A. Hirst: "On the correlation of two planes". Proc. London Math. Soc. 5 (1874), 40-70.

[7] T.A. Hirst: "On correlations in space". Proc. London Math. Soc. 6 (1874), 7-9.

[8] T.A. Hirst: "On the correlation of two planes". Annali di matematica 6 (1875), 260-297.

[9] T.A. Hirst: "Note on the correlations of two planes". Proc. London Math. Soc. 8 (1877), 262-273.

[10] T.A. Hirst: "Note on the correlation of two planes". Annali de matematica pura ed applicata 8 (1878), 287-300.

[11] T.A. Hirst: "On the correlation of two spaces each of three dimensions". Proc. London Math. Soc. 21 (1890), 92-118.

[12] G. Loria: "Sulle corrispondenze projettive fra due piani e fra due spazii". Giorn. di Math. 22 (1884), 1-16.

[13] Predella: "Le omografie in uno spazio ad un numero qualunque di dimensioni". Ann. di mat. 17 (1890), 113-159.

[14] H. Schubert: "Die n-dimensionalen Verallgemeinerungen der fundamentalen Anzahlen unseres Raums". Math. Ann. 26 (1886), 26-51.

[15] H. Schubert: "Über Räume zweiten Grades". Mitteil. der Hamb. Math. ges. 1 (1889), 290-310.

[16] H. Schubert: "Kegelschnitt-Anzahlen als Functionen der Raum-Dimension". Mitteil. der Hamb. Math. Ges. 2 (1890), 172-184.

[17] H. Schubert: "Über eine Verallgemeinerung der Aufgaben der abzählenden Geometrie". Mitteil. der Hamb. Math. Ges. 3 (1891), 12-20.

[18] H. Schubert: "Beitrag zur Liniengeometrie in n Dimensionen". Mitteil. der Hamb. Math. Ges. 3 (1891), 86-97.

[19] H. Schubert: "Beziehungen zwischen den linearen Räumen auferlegbaren characteristischen Bedingungen". Math. Ann. 38 (1891), 598-602.

[20] H. Schubert: "Allgemeine Anzahlfunctionen für Kegelschnitte, Flächen und Räume zweiten Grades in n Dimensionen". Math. Ann. 45 (1894), 153-206.

[21] H. Schubert: "Correlative Verwandtschaft in n Dimensionen". Jaresher. der Deutsch. Math. Verein 4 (1894-95).

[22] C. Segre: "Sulla teoria e sulla classificazione delle homografie in uno spazio lineare ad un numero qualunque di dimensioni". Memorie dell'Acc. dei Lincei 19 (1883-84), 127-148.

[23] C. Segre: "Studio sulle quadriche in uno spazio lineare ad un numero qualunque di dimensioni". Mem. Acc. Torino 36 (1884), 3.

[24] J.G. Semple: "On complete quadrics (I)". Journal London Math. Soc. 23 (1984), 258-267.

[25] J.G. Semple: "The variety whose points represent complete collineations of S_r on S_r'". R.C. Math. Univ. Roma 10 (1951), 201-207.

[26] J.G. Semple: "On complete quadrics (II)". Journal London Math. Soc. 27 (1952), 280-287.

[27] F. Severi: "Sui fondamenti della geometria numerativa e sulla teoria delle caratteristiche". Atti del R. Inst. Veneto 75 (1916), 1122-1162.

[28] F. Severi: "I fondamenti della geometria numerativa". Ann. di Mat. 19 (1940), 151-242.

[29] E. Study: "Über die Geometrie der Kegelschnitte, insbesondere deren Characteristikenproblem". Math. Ann. 26 (1886), 58-101.

[30] R. Sturm: "Das Problem der Projectivität und seine Anwendung auf Die Flächen zweiten Grades". Math. Ann. 1 (1869), 533-574.

[31] R. Sturm: "Über correlative oder reciproke Bündel". Math. Ann. 12 (1877), 254-368.

[32] R. Sturm: Über die resiproke und mit ihr zusammenhängende Verwandtschaften". Math. Ann. 19 (1882), 461-488.

[33] J.A. Tyrrell: "Complete quadrics and collineations in S_n". Mathematika 3 (1956), 69-79.

[34] I. Vainsencher: "Schubert calculus for complete quadrics". To appear in the Proceedings of the Nice conference on enumerative geometry Summer 1981. Birkhäusen.

[35] P. Visalli: "Sulle correlazioni in due spazi a tre dimensioni". Memorie de l'Acc. dei Lincei (4) 3 (1886), 597-671.

[36] P. Visalli: "Sulle collinearita e correlazioni ordinarie ed eccezionali a quattro dimensioni. 1, 2, 3, 3 cont.". Rend. Inst. Lombardo 29 (1986), 351-359, 439-459, 521-528, 559-565.

[37] P. Visalli: "Sulle correlazioni in due spazi a tre dimensioni". Rendiconti della R. Accademia dei Lincei (4) 3 (1887).

[38] B.L. van der Waerden: "Z.A.G. XV. Lösung des Characteristikenproblems für Kegelschnitte". Math. Ann. 115 (1938), 645-655.

Department of Mathematics
University of Stockholm
Box 6701

S 11385 STOCKHOLM
(Sweden)

DIVISEURS SPECIAUX ET INTERSECTION DE CYCLES DANS
LA JACOBIENNE D'UNE COURBE ALGEBRIQUE

Arnaud BEAUVILLE

INTRODUCTION.

Dans les démonstrations de géométrie énumérative, le point clé est souvent le calcul des classes de cohomologie de certains cycles algébriques. Il n'est guère plus difficile de travailler, non pas dans la cohomologie, mais dans l'anneau de Chow des variétés considérées ; on obtient ainsi, outre le résultat numérique, une information géométrique qui peut être intéressante.

Dans ce travail, nous appliquons cette philosophie à la formule de Castelnuovo qui calcule le nombre de systèmes linéaires spéciaux sur une courbe générique, lorsque ce nombre est fini. Le résultat est la partie c) du théorème suivant.

THEOREME 1. *Soit C une courbe générale de genre g, et soient r, d deux entiers tels que* $g = (r+1)(g+r-d)$.

a) Il existe un nombre fini $N(g,r,d)$ *de classes de diviseurs* D_i *sur C, de degré d, telles que* $h^0(D_i) = r+1$.

b) On a $N(g,r,d) = g!\,\dfrac{1!\ldots\ldots r!}{(g+r-d)!\ldots(g+2r-d)!}$

c) La somme dans Pic(C) de ces diviseurs est un multiple entier du diviseur canonique.

On a donc $\sum_{i=1}^{N} D_i \equiv mK$, où $m = \dfrac{dN(g,r,d)}{2g-2}$ est entier.

La partie a) de ce théorème est un cas particulier du problème de Brill-Noether, récemment résolu par Griffiths-Harris [G-H] . La formule b) est due à Castelnuovo [C] . L'assertion c) m'a été signalée par E. Arbarello, comme conséquence de la conjecture suivante :

CONJECTURE (Franchetta) : *Soient* $\mathcal{M}_g$ *l'espace des modules (grossier) des courbes de genre g* ($g \geq 2$), M_g *son corps de fonctions, et* C_g *la courbe universelle sur* M_g. *Le groupe* $\mathrm{Pic}(C_g)$ *est engendré par la classe du diviseur canonique.*

La démonstration proposée par Franchetta [F] est malheureusement tout-à-fait incomplète (un gros "trou" figure au début du n°11, p.140), et on ne voit guère comment la sauver. C'est pourquoi il m'a paru intéressant de donner de l'assertion c) une démonstration directe, proche de la démonstration de la formule énumérative b). (*)

§ I. VARIETES DE DIVISEURS SPECIAUX DANS LES JACOBIENNES.

Toutes les variétés considérées dans cet article sont des variétés complexes. Si X est une variété lisse projective, on notera $CH(X) = \bigoplus_p CH^p(X)$ l'anneau de Chow de X (anneau des cycles algébriques modulo l'équivalence rationnelle, gradué par la codimension).

On fixe désormais une courbe algébrique C lisse, connexe, de genre g. Pour $d \in \mathbb{Z}$, on note J^d la variété des classes de diviseurs de degré d sur C (modulo équivalence linéaire) ; ainsi $J = J^o$ est la jacobienne de C, et J^d est isomorphe (non canoniquement) à J. On désigne par W^r_d la sous-variété de J^d formée des classes des diviseurs D tels que $h^o(D) \geq r+1$.

On note $C^{(d)}$ la d-ième puissance symétrique de C et $\varphi^d : C^{(d)} \longrightarrow J^d$ le morphisme canonique. Rappelons que φ^d est génériquement injectif pour $0 \leq d \leq g$; son image, qui n'est autre que W^o_d, est donc de dimension d.

Il sera commode dans ce qui suit d'identifier entre elles les variétés J^d ; on choisit pour cela un point P de C, et on identifie désormais J^d à J à l'aide de la translation par le diviseur dP. Avec cette convention, on note $w_p \in CH^p(J)$ la classe du cycle W^o_{g-p}. On notera aussi θ l'élément w_1 de Pic(J) : il définit la polarisation principale de J (théorème de Riemann).

Le théorème de Kleiman-Laksov [K-L] exprime le cycle W^r_d en fonction des w_p :

THEOREME 2. *Soient d,r des entiers positifs ; posons* $\rho = g-(r+1)(g-d+r)$.

a) Les composantes de W^r_d *sont de dimension* $\geq \rho$ *; il y a égalité si C est générique.*

b) Si la variété W^r_d *est pure de dimension* ρ *, sa classe dans* $CH^{g-\rho}(J)$ *est*

(*) La conjecture est maintenant démontrée, comme conséquence de résultats topologiques de J. Harer.

$$[W_d^r] = \det \begin{vmatrix} w_{g+r-d} & \cdots\cdots & w_{g+2r-d} \\ \vdots & & \\ w_{g-d} & \cdots\cdots & w_{g+r-d} \end{vmatrix} .$$

COROLLAIRE. *Sous l'hypothèse b), la classe de W_d^r dans $H^{2g-2\rho}(J,\mathbb{Z})$ est* $\theta^{g-\rho} \dfrac{1!\ldots\ldots r!}{(g+r-d)!\ldots(g+2r-d)!}$.

Cela résulte de la formule de Poincaré : $w_p = \dfrac{\theta^p}{p!}$ dans $H^{2p}(J,\mathbb{Z})$ et d'un calcul facile de déterminant, cf. [K-L] .

On déduit aussitôt du corollaire la formule b) du théorème 1 (noter que $\deg\theta^g = g!$).

Pour obtenir le résultat plus précis c), il faut calculer dans CH(J), modulo une relation d'équivalence plus fine que l'équivalence homologique. Observons que pour toute variété abélienne A de dimension g, l'application qui associe à un 0-cycle $\Sigma m_i[a_i]$ ($m_i \in \mathbb{Z}$, $a_i \in A$) l'élément $\Sigma m_i\, a_i$ de A définit par passage au quotient un homomorphisme $S : CH^g(A) \longrightarrow A$ (cf. [W]). Prenons A=J et plaçons-nous sous les hypothèses du théorème 1, de sorte que W_d^r est un ensemble fini de diviseurs $D_1,\ldots,D_N$; on trouve alors $S(W_d^r) = \Sigma\, D_i - (Nd)P$. Pour déduire le théorème 1 c) du théorème 2, il s'agit donc de calculer les expressions $S(w_{p_1}\ldots w_{p_n})$ pour $p_1+\ldots+p_n = g$.

<u>THEOREME 3</u>. *Soient $p_1,\ldots,p_n$ des entiers positifs de somme g.* *On a : $S(w_{p_1}\ldots w_{p_n}) = M(g;p_1,\ldots,p_n)\,(K - (2g-2)P)$,*

où $M(g;p_1,\ldots,p_n)$ est un entier défini par

$$M(g;p_1,\ldots,p_n) = \frac{(g-2)!}{p_1!\ldots p_n!}\,\frac{1}{2}\sum_{i=1}^{n} p_i(g-p_i)$$

Ce théorème, joint au théorème 2, entraîne qu'il existe un entier m tel que $\Sigma D_i - (Nd)P \equiv m(K-(2g-2)P)$. Cette formule étant valable pour tout point P de C, on en déduit $m = \dfrac{Nd}{2g-2}$ et $\Sigma\, D_i \equiv mK$, d'où le théorème 1 c). On notera qu'il n'est pas nécessaire pour cela de connaître la valeur de l'entier $M(g;p_1,\ldots,p_n)$.

<u>REMARQUE</u>. Il résulte de la démonstration que le théorème 1 c) est valable pour une courbe C quelconque, pourvu que la variété W_d^r soit finie (et $\rho = 0$). Sans cette hypothèse, W. Fulton m'a indiqué qu'il existe toujours un 0-cycle effectif $\mathfrak{z}$ de W_d^r, bien défini à équivalence rationnelle près, dont la classe dans $CH^g(J)$ est donnée par la formule

du th. 2 b) : on peut l'obtenir par exemple en écrivant C comme spécialisation d'une courbe générale. La somme des diviseurs de $\mathfrak{z}$ est alors un multiple entier du diviseur canonique.

§ II. CYCLES ALGEBRIQUES SUR LES VARIETES ABELIENNES.

Je regroupe dans ce paragraphe quelques résultats généraux utiles pour la suite. Je renvoie par exemple à [G] pour la définition des jacobiennes intermédiaires et de l'application d'Abel-Jacobi.

PROPOSITION 1. *Soit* X *une variété projective lisse. Pour* $1 \leq p \leq \dim X$, *notons* $\Sigma^p(X)$ *le sous-groupe de* $CH^p(X)$ *formé des cycles homologiquement équivalents à zéro,* $J^p(X)$ *la p-ième jacobienne intermédiaire de* X *et* $\alpha^p : \Sigma^p(X) \longrightarrow J^p(X)$ *l'application d'Abel-Jacobi. Soient* $x \in \Sigma^p(X)$, $y \in \Sigma^q(X)$; *on a alors* $\alpha^{p+q}(xy) = 0$

DEMONSTRATION : Pour tout cycle $y \in CH^q(X)$, on a un diagramme commutatif

$$\begin{array}{ccc} \Sigma^p(X) & \xrightarrow{\ y\ } & \Sigma^{p+q}(X) \\ {\scriptstyle \alpha^p}\downarrow & & \downarrow{\scriptstyle \alpha^{p+q}} \\ J^p(X) & \longrightarrow & J^{p+q}(X) \end{array}$$

où la première flèche horizontale est l'intersection avec y, et la seconde est déduite du cup-produit avec la classe de y dans $H^{2q}(X, \mathbb{Z})$ ([G], 2.16). Lorsque y est homologiquement trivial, on a donc $\alpha^{p+q}(x\ y) = 0$ pour tout x dans $\Sigma^p(X)$.

Lorsque $p = \dim(X)$, le tore $J^p(X)$ est la variété d'Albanese de X, et α^p est l'homomorphisme canonique déduit du morphisme d'Albanese. Si X est elle-même une variété abélienne, Alb(X) est canoniquement isomorphe à X, et $\alpha^p : \Sigma^p(X) \longrightarrow X$ n'est autre que la restriction à $\Sigma^p(X)$ de l'homomorphisme S défini au n° 2. Par conséquent :

PROPOSITION 2. *Soit* A *une variété abélienne, et soient* x, y *deux cycles sur* A *de dimensions complémentaires, homologiquement équivalents à zéro. On a* $S(xy) = 0$.

Les corollaires qui suivent se trouvent déjà dans [M] . Pour $a \in A$, on note T_a la translation $x \longmapsto x+a$ de A.

COROLLAIRE 1. *Soient x,y des cycles de dimensions complémentaires sur A. L'application $\alpha(x,y)$: $a \longmapsto S(x.(T^*_a y - y))$ est un endomorphisme de A, qui ne dépend que des classes de cohomologie de x et y.*

DEMONSTRATION : Notons d'abord que pour tout 0-cycle $\mathfrak{z} \in CH^g(A)$, on a

$$S(T^*_a \mathfrak{z}) = S(\mathfrak{z}) - (\deg \mathfrak{z})\, a$$

par conséquent : $S(x.T^*_a y) = S(T^*_{-a} x.y) - (\deg x.y)\, a$

et $\quad S(x.(T^*_a y - y)) = S((T^*_{-a} x - x).y) - (\deg x.y)\, a \quad .$

Il résulte alors de la prop.2 que cette expression est nulle lorsque x ou y est homologiquement trivial. Remplaçant y par $T^*_b y - y$, pour $b \in A$, on en déduit

$$S(x.(T^*_{a+b} y - T^*_a y - T^*_b y + y)) = 0,$$

ce qui prouve que $\alpha(x,y)$ est un homomorphisme.

COROLLAIRE 2. *Posons $g = \dim(A)$. Soient $p_1,\dots,p_n$ des entiers positifs de somme g, et soient*

$$x_1 \in CH^{p_1}(A),\dots,x_n \in CH^{p_n}(A),\ a_1,\dots,a_n \in A.$$

a) *On a*

$$S(T^*_{a_1} x_1 \cdot \dots \cdot T^*_{a_n} x_n) - S(x_1 \dots x_n) = \sum_i \alpha(x_1 \dots \hat{x}_i \dots x_n, x_i)(a_i)$$

b) *Soit D un diviseur sur A ; supposons que pour $1 \le i \le n$, x_i soit homologiquement équivalent à D^{p_i}. On a alors*

$$S(T^*_{a_1} x_1 \cdot \dots \cdot T^*_{a_n} x_n) - S(x_1 \dots x_n) = -\frac{\deg D^g}{g} \sum_i p_i\, a_i$$

La formule a) résulte du cor.1 et de l'égalité

$$S(T^*_{a_1} x_1 \cdot \dots \cdot T^*_{a_n} x_n) = S((T^*_{a_1} x_1 - x_1) \cdot \dots \cdot T^*_{a_n} x_n) +$$

$$+ S(x_1.(T^*_{a_2} x_2 - x_2) \dots T^*_{a_n} x_n) + S(x_1 \dots x_{n-1}(T^*_{a_n} x_n - x_n)) + S(x_1 \dots x_n).$$

En faisant dans la formule a) $x_i = D$ pour tout i, $a_1 = \dots = a_p = 0$, $a_{p+1} = \dots = a_g = a$, on obtient

$$\alpha(D^p, D^{g-p}) = (g-p)\ \alpha(D^{g-1}, D) \ ;$$

en particulier, pour p = 0, on trouve

$$\alpha(D^{g-1}, D) = \frac{1}{g}\, \alpha(1, D^g) = -\frac{1}{g} (\deg D^g).\, Id_A$$

d'où finalement $\alpha(D^p, D^{g-p}) = \frac{p-g}{g} (\deg D^g).\, Id_A$

En reportant cette égalité dans a), on obtient la formule b).

§ 3. DEMONSTRATION DU THEOREME 3.

a) Réduction à un calcul modulo torsion

Nous allons d'abord nous réduire à un résultat moins fort :

PROPOSITION 3. *Il existe un rationnel* M *tel qu'on ait* $S(w_{p_1}\ldots w_{p_n}) = M(K-(2g-2)P)$ *dans* $J \otimes \mathbb{Q}$.

Montrons que la prop. 3 entraîne le théorème 3. Examinons pour cela l'effet d'un changement du point de base P. Soit P' un autre point de C ; posons $\alpha = P'-P$. La sous-variété W'_q de J définie comme au § 1 à l'aide du point P' est égale à $T^{-1}_{q\alpha}(W_p)$; sa classe w'_p dans $CH^p(J)$ (avec $p = g-q$) est donc $T^*_{(g-p)\alpha}\, w_p$. D'après le cor. 2b) du § 2, on a

$$S(w'_{p_1}\ldots w'_{p_n}) - S(w_{p_1}\ldots w_{p_n}) = -\frac{(g-1)!}{p_1!\ldots p_n!}\sum_i p_i\,(g-p_i).\alpha \ ;$$

d'autre part en appliquant la prop.3, on trouve que cette expression est égale à $-M(2g-2)\alpha$, d'où $M = \frac{(g-2)!}{p_1!\ldots p_n!}\cdot\frac{1}{2}\sum p_i(g-p_i) = M(g;p_1,\ldots,p_n)$.

LEMME 1. *Le nombre M est entier.*

DEMONSTRATION. Ecrivant $p_i(g-p_i) = p_i(g-1) - p_i(p_i-1)$, on obtient

$$2M = \frac{g!}{p_1!\ldots p_n!} - \sum_i \frac{(g-2)!}{p_1!\ldots(p_i-2)!\ldots p_n!}.$$

Par conséquent, 2M est le coefficient de $T^{p_1}\ldots T^{p_n}$ dans le développement de

$$(T_1+\ldots+T_n)^g - (T_1+\ldots+T_n)^{g-2}(T_1^2+\ldots+T_n^2) = 2(T_1+\ldots+T_n)^{g-2}\sum_{i<j} T_i\,T_j\ ,$$

d'où le lemme.

Il reste à éliminer la torsion. Soit $f:\mathcal{C}\longrightarrow S$ une famille de courbes lisses ; l'application $P\longmapsto S(w_{p_1}\ldots w_{p_n}) - M(K-(2g-2)P)$ définit un morphisme de $\mathcal{C}$ dans le schéma de Picard Pic $(\mathcal{C}/S)$. D'après ce qui précède, ce morphisme est constant sur les fibres de f, et son image est de torsion ; pour un entier ℓ convenable, il définit donc une section du revêtement $S_\ell \longrightarrow S$ formé des points d'ordre ℓ de Pic $(\mathcal{C}/S)$.

Prenons maintenant pour S le schéma de Hilbert des courbes de genre g tricanoniques dans $\mathbb{P}^{5g-6}$ (de sorte que le quotient $S/PGL(5g-5)$

est l'espace des modules $\mathcal{M}_g$), et pour $\mathcal{C}$ la courbe universelle. Alors S_ℓ est irréductible : en effet $S_\ell/PGL(5g-5)$ est un quotient de l'espace de Teichmüller. Le revêtement $S_\ell \longrightarrow S$ ne peut donc avoir une section que pour $\ell=1$, ce qui prouve l'égalité $S(w_{p_1}\ldots w_{p_n}) = M\,(K-(2g-2)P)$ dans $J(C)$ pour toute courbe C et tout point $P \in C$.

b) Réduction à un calcul sur C^p.

Nous calculons désormais dans $J \otimes \mathbb{Q}$. On a

$$S(w_{p_1}\ldots w_{p_n}) = S((w_{p_1} - \frac{\theta^{p_1}}{p_1!})\ldots w_{p_n}) + S(\frac{\theta^{p_1}}{p_1!}(w_{p_2} - \frac{\theta^{p_2}}{p_2!})\ldots w_{p_n}) +$$

$$+ S(\frac{\theta^{p_1}}{p_1!}\ldots(w_{p_n} - \frac{\theta^{p_n}}{p_n!})) + S(\frac{\theta^{p_1}}{p_1!}\ldots\frac{\theta^{p_n}}{p_n!})$$

Le cycle $w_{p_i} - \frac{\theta^{p_i}}{p_i!}$ est homologiquement équivalent à zéro ; compte tenu de la prop.2, on trouve :

$$p_1!\ldots p_n!\,S(w_{p_1}\ldots w_{p_n}) = \sum_{i=1}^{n} S(\theta^{g-p_i}.(p_i!w_{p_i} - \theta^{p_i})) + S(\theta^g).$$

Pour démontrer la prop.3, il suffit donc de prouver que pour tout p, l'élément $S(w_{g-p}\,\theta^p)$ de $J \otimes \mathbb{Q}$ est multiple rationnel de $K-(2g-2)P$.

Fixons un entier positif $p \le g$; notons simplement φ l'application φ^p de $C^{(p)}$ dans J. Soient $\pi : C^p \longrightarrow C^{(p)}$ l'application canonique, et $\rho = \varphi \circ \pi$. On a dans $CH^g(J) \otimes \mathbb{Q}$

$$\begin{aligned} w_{g-p}\,\theta^p &= (\varphi_* 1).\,\theta^p = \varphi_*\,\varphi^*\,\theta^p \\ &= \frac{1}{p!}\,\varphi_*\pi_*\pi^*\varphi^*\,\theta^p = \frac{1}{p!}\,\rho_*(\rho^*\theta)^p\,. \end{aligned}$$

Reste à calculer $\rho^*\theta$. On notera $p_i : C^p \longrightarrow C$ la i-ième projection, et Δ_{ij} le diviseur de C^p formé des points $(x_1,\ldots,x_p)$ tels que $x_i=x_j$.

LEMME 2. *Posons $D = K - (g-1-p)P$. On a dans $\mathrm{Pic}(C^p)$ l'égalité*

$$\rho^*\theta = \sum_i p_i^*\,D - \sum_{i<j} \Delta_{ij}\,.$$

DEMONSTRATION : Les deux expressions sont invariantes sous le groupe symétrique $\mathfrak{S}_p$; d'après les théorèmes du carré et du cube, il suffit de vérifier l'égalité du lemme après restriction à $C \times \{x_1\} \times \ldots \times \{x_{p-1}\}$, pour tous points $x_1,\ldots,x_{p-1}$ de C. Posant $\lambda = x_1+\ldots+x_{p-1} - (p-1)P$, on est ramené à démontrer dans Pic(C) l'égalité

$$(\varphi^1)^* T_\lambda^* \theta = D - \sum_i x_i = K - (g-2)P - \lambda ,$$

qui est bien connue ([W], n° 41, th. 20), et facile : il suffit de la vérifier pour $\lambda = x-y$ $(x,y \in C)$, auquel cas elle résulte aussitôt de Riemann-Roch.

c) Le calcul.

Il s'agit donc de prouver que chaque terme du développement de $\rho_*(\sum_i p_i^* D - \sum_{i<j} \Delta_{ij})^p$ est multiple rationnel (dans $J \otimes \mathbb{Q}$) de $K-(2g-2)P$. Nous décrirons un tel terme par un graphe $\Gamma = (A,S)$, c'est-à-dire un ensemble d'arêtes A, un ensemble de sommets S et une application e de A dans l'ensemble des parties de S formées de un ou deux éléments. On prendra ici $S = [1,p]$; à un tel graphe on associe le cycle

$$t(\Gamma) = \prod_{a \in A} \Delta_{e(a)} ,$$

en posant par convention $\Delta_s = p_s^* D$ pour $s \in S$; on a $t(\Gamma) \in CH^q(C^p)$, avec $q = \mathrm{Card}(A)$. Il est clair que $(\sum_i p_i^* D - \sum_{i<j} \Delta_{ij})^p$ est combinaison linéaire à coefficients entiers de cycles $t(\Gamma)$, pour des graphes Γ ayant p arêtes.

Nous fixons dans ce qui suit un graphe Γ avec $\mathrm{Card}(A) = p$, tel que $t(\Gamma)$ soit non nul. Pour toute partie I de S, on note p_I la projection de C^S sur C^I, et $\delta_I : C \longrightarrow C^I$ l'application diagonale.

LEMME 3. *Pour toute composante connexe* $\Gamma_\alpha = (A_\alpha, S_\alpha)$ *de* Γ *, on a* $\mathrm{Card}(A_\alpha) = \mathrm{Card}(S_\alpha)$.

DEMONSTRATION : Pour $a \in A_\alpha$, le diviseur $\Delta_{e(a)}$ est l'image réciproque par la projection p_{S_α} d'un diviseur sur C^{S_α} ; par conséquent le cycle $\prod_{a \in A_\alpha} \Delta_{e(a)}$ est l'image réciproque d'un cycle non nul sur C^{S_α}, de codimension $\mathrm{Card}(A_\alpha)$. On a donc $\mathrm{Card}(A_\alpha) \leq \dim(C^{S_\alpha}) = \mathrm{Card}(S_\alpha)$.
Comme

$$p = \sum_\alpha \mathrm{Card}(A_\alpha) \leq \sum_\alpha \mathrm{Card}(S_\alpha) = p,$$

on a $\mathrm{Card}(A_\alpha) = \mathrm{Card}(S_\alpha)$ pour tout α .

Soit n un entier ≥ 1. On dit qu'un graphe $\Gamma = (A,S)$ est un <u>circuit de longueur n</u> si on peut écrire $A = \{a_1,\dots,a_n\}$, $S = \{s_1,\dots,s_n\}$ de façon que $e(a_i) = \{s_i,s_{i+1}\}$ pour tout $1 \leq i \leq n-1$, et $e(a_n) = \{s_n,s_1\}$.

<u>LEMME 4.</u> *Soit $\Gamma' = (B,T)$ un circuit de longueur n. On a $t(\Gamma') = p_T^*(\delta_T)_* E_n$, avec $E_n = -K$ si $n \geq 2$ et $E_1 = D$.*

<u>DEMONSTRATION</u> : Le cas n=1 étant évident, on peut supposer $n \geq 2$; il s'agit de prouver dans $CH^n(C^n)$ l'égalité $\Delta_{12}\dots\Delta_{n-1,n}\,\Delta_{n,1} = -\delta_* K$, où $\delta : C \longrightarrow C^n$ est l'application diagonale.

On a

$$\Delta_{12}\ \dots\ \Delta_{n-1,n} = \delta_* 1 \text{ et } \delta_* 1 \,.\, \Delta_{n,1} = \delta_*(\delta^* \Delta_{n,1}) \ .$$

Or $\Delta_{n,1}$ est l'image réciproque par $p_{n,1} : C^n \longrightarrow C^2$ de la diagonale de CxC ; par suite $\delta^* \Delta_{n,1}$ est l'autointersection de la diagonale dans CxC, c'est-à-dire -K.

<u>LEMME 5.</u> *Soient $I \subset S$, $i \in I$, $j \in S-I$ et $J = I \cup \{j\}$. Pour tout diviseur E sur C, on a*

$$(p_I^*(\delta_I)_* E) \,.\, \Delta_{ij} = p_J^*(\delta_J)_* E$$

Il suffit de le démontrer lorsque E est réduit à un point ; la vérification est alors immédiate.

Nous pouvons maintenant calculer $t(\Gamma)$. Soit $\Gamma_\alpha = (A_\alpha,S_\alpha)$ une composante de Γ . C'est un graphe connexe de genre 1, c'est-à-dire réunion d'un circuit et d'arbres disjoints, rencontrant chacun le circuit en un point. Il résulte aussitôt des lemmes 4 et 5 qu'on a $t(\Gamma_\alpha) = p_{S_\alpha}^* (\delta_{S_\alpha})_* E_n$. Par conséquent

$$t(\Gamma) = \prod_\alpha p_{S_\alpha}^* (\delta_{S_\alpha})_* E_\alpha ,$$

où E_α est égal à D si le graphe Γ_α contient une arête à une seule extrémité et à -K dans le cas contraire.

On a alors

$$\rho_* t(\Gamma) = \sum_\alpha m_\alpha (E_\alpha - (\deg E_\alpha).P), \text{ avec } m_\alpha = \mathrm{Card}(S_\alpha) \,.\, \prod_{\beta\neq\alpha} \deg(E_\beta) \ ;$$

en effet, pour vérifier cette formule, on peut supposer que tous les E_α sont réduits à un point, auquel cas c'est immédiat. Compte tenu de la valeur des E_α, on trouve

$$\rho_* \, t(\Gamma) = M\,(K - (2g-2)P, \text{ avec } M = \Sigma \pm m_\alpha \in \mathbb{Z}.$$

Ceci achève de prouver la proposition 3, donc le théorème 3.

BIBLIOGRAPHIE

[C] G. Castelnuovo : Numero delle involuzioni giacenti sopra una curva di dato genere. Rend. Acad. Lincei, s.4, 5 (1889).

[F] A. Franchetta : Sulle serie lineari razionalmente determinate sulla curva a moduli generali di dato genere. Le Matematiche, Fasc. 11 (1954).

[G] P. Griffiths : Some transcendental methods in the study of algebraic cycles. Several complex variables II, Maryland 1970 - Springer Lecture Notes 185 (1971), 1-46.

[G-H] P. Griffiths, J. Harris : On the variety of special linear systems on a general algebraic curve. Duke Math. Journal 47 (1980), 233-272.

[K-L] S. Kleiman, D. Laksov : Another proof of the existence of special divisors. Acta Math. 132 (1974), 163-176.

[M] T. Matsusaka : On a characterization of a Jacobian variety. Mem. Coll. Sci. Kyoto, ser. A, (1959), 1-19.

[W] A. Weil : Courbes algébriques et variétés abéliennes. Hermann, Paris (1948).

Arnaud BEAUVILLE
Centre de Mathématiques
Ecole Polytechnique

F-91128 PALAISEAU CEDEX.

FIBRES DE 't HOOFT SPECIAUX ET APPLICATIONS

A. HIRSCHOWITZ et M.S. NARASIMHAN

On s'intéresse aux fibrés vectoriels algébriques de rang deux sur un espace projectif de dimension deux ou trois sur un corps de base algébriquement clos.

Les fibrés de 't Hooft ([6] § 3.1.1) sont les fibrés de rang deux sur $\mathbb{P}^3$ associés aux réunions disjointes de droites. Nous introduisons les fibrés de 't Hooft spéciaux ('tHS) : ce sont les fibrés associés aux réunions de droites situées sur une même quadrique non-singulière. L'étude de ces fibrés nous conduit aux résultats suivants concernant l'espace des modules $M(c_1,c_2)$ des fibrés stables (de rang deux) de classes de Chern c_1 et c_2 :

THÉORÈME 1. *L'espace $M(0,2)$ des modules de fibrés instantons sur $\mathbb{P}^3$ avec $c_1 = 0$, $c_2 = 2$ est rationnel.*

THÉORÈME 2. *Si k est pair il n'existe pas de fibré universel algébrique (fibré de Poincaré) sur $\mathbb{P}^3 \times M_{\mathbb{P}^3}(0,k)$.*

Pour simplifier l'écriture, nous normalisons nos fibrés par $c_1=2$. Rappelons que $M(0,k)$ est isomorphe à $M(2,k+1)$. On sait ([6]) que tout fibré E de $M(2,3)$ s'insère dans une suite exacte S(E) :

$$0 \longrightarrow \mathcal{O}\oplus\mathcal{O} \longrightarrow E \longrightarrow \tilde{\mathcal{F}}_E \longrightarrow 0$$

où $\tilde{\mathcal{F}}_E$ est le prolongement par zéro d'un fibré $\mathcal{F}_E$ de rang 1 et bidegré $\{2,-1\}$ sur une quadrique non singulière Q_E. Le faisceau $\tilde{\mathcal{G}}_E := \mathcal{E}xt^1(\tilde{\mathcal{F}}_E,\mathcal{O})$ est le prolongement par zéro d'un fibré $\mathcal{G}_E$ de rang 1 et bidegré $\{0,3\}$ sur Q_E et le foncteur $\mathcal{H}om(.,\mathcal{O})$ appliqué à S(E) fournit deux sections qui engendrent $\mathcal{G}_E$ et donc un pinceau sans point de base de diviseurs de bidegré $\{0,3\}$ sur Q_E. Cette construction permet d'identifier M comme variété des pinceaux sans point de base de bidegré $\{0,3\}$ sur une quadrique non singulière. Nous donnons au § 4 une démonstration détaillée de cette identification qui fait de M(2,3) un fibré sur la variété $\tilde{\mathscr{K}}$ des quadriques non singulières de $\mathbb{P}^3$ munies d'un système de génératrices. Revêtement à deux feuillets de la variété $\mathscr{K}$ des quadriques non singulières, $\tilde{\mathscr{K}}$ est une variété rationnelle (cf § 3), et si C est le fibré en courbes rationnelles tautologique sur $\tilde{\mathscr{K}}$, alors M(2,3) est le fibré sur $\tilde{\mathscr{K}}$ des pinceaux sans point de base de degré 3 sur les fibres de C. C'est donc un ouvert d'un fibré en variétés grassmanniennes. Ce fibré en grassmanniennes n'est pas localement trivial dans la topologie de Zariski, mais il admet un sous-fibré ouvert localement trivial dans cette topologie (cf §2). Par suite M(2,3) est rationnel et p: M(2,3) $\longrightarrow \tilde{\mathscr{K}}$ admet des sections rationnelles. La suite exacte duale de S(E) :

$$0 \longrightarrow E^{v} \longrightarrow \mathcal{O} \oplus \mathcal{O} \longrightarrow \tilde{\mathscr{G}}_E \longrightarrow 0$$

permet de voir que s'il existe un fibré universel, alors $p^{*}C \longrightarrow M(2,3)$ provient d'un fibré vectoriel (Prop. 4.8). Ce n'est pas le cas parce que $C \longrightarrow \tilde{\mathscr{K}}$ ne provient pas d'un fibré vectoriel (§3) et que p admet une section rationnelle.

Les fibrés 'tHS sont les fibrés de M(2,2k+1) auxquels les raisonnements précédents s'étendent sans changement. Ils permettent de prouver le théorème 2.

La rationalité des modules de fibrés stables est connue dans certains cas sur les courbes ([11]) et sur $\mathbb{P}^2$ ([1] [4] [7]). Quant à la question du fibré universel, elle a d'abord été étudiée sur les courbes ([10] [15]). Maruyama [9] a ensuite donné en général une condition suffisante d'existence ; dans les cas qui nous préoccupent ici, cette condition est vérifiée pour $c_1=0$ et c_2 impair. Enfin Le Potier [8] a prouvé que pour les fibrés de rang deux sur $\mathbb{P}^2$, cette condition est aussi nécessaire. Nous expliquons rapidement au § 5 comment notre méthode permet de retrouver ce résultat de Le Potier dans le cas algébrique. Signalons aussi que Newstead [13] a étudié plus spécialement $M_{\mathbb{P}^3}(0,2)$: la lecture de [13] est d'ailleurs à l'origine du présent travail.

Le Potier a démontré en même temps que nous le théorème 1 et le théorème 2 dans le cas particulier où k=2. Inspiré par les idées de Le Potier, Newstead démontre dans [14] les résultats de notre § 2.

1. FIBRÉS PROJECTIFS ET FIBRÉS EN CONIQUES.

Dans ce qui suit, un fibré projectif est un morphisme $\pi : C \longrightarrow B$ propre et plat de variétés algébriques (irréductibles) lisses dont les fibres sont isomorphe à l'espace projectif $\mathbb{P}^m$ de dimension m. Quand m=1, les fibres sont des courbes rationnelles lisses et on dit que π est un fibré en coniques.
Le résultat suivant est bien connu [17, Prop.18]

PROPOSITION 1.1. *Soit $\pi: C \longrightarrow B$ un fibré projectif. Les conditions suivantes sont équivalentes :*

1) Le fibré projectif est isomorphe au fibré projectif associé à un fibré vectoriel sur B.

2) $\pi : C \longrightarrow B$ admet une section rationnelle.

3) Il existe un fibré en droites sur C qui vaut $\mathcal{O}(1)$ sur chaque fibre.

On dira qu'un fibré vérifiant ces conditions est banal.

REMARQUE 1.2. Supposons que S est une sous-variété lisse de B tel que $\pi^{-1}(S) \longrightarrow S$ ne soit pas banal. Alors si U est un ouvert non-vide de B, $\pi^{-1}(U) \longrightarrow U$ n'est pas banal. S'il l'était, $\pi : C \longrightarrow B$ serait banal par la proposition 1.1. et donc $\pi^{-1}(S) \longrightarrow S$ aussi.

REMARQUE 1.3. Soit $\pi : C \longrightarrow B$ un fibré en coniques. Le fibré tangent relatif T_π est un fibré en droites qui est de degré 2 sur chaque fibre. Il s'ensuit que le fibré π est banal s'il existe un fibré en droites sur C qui est de degré k sur chaque fibre, avec k impair.

Nous avons le résultat suivant assurant qu'un fibré en coniques n'est pas banal. Ce résultat remonte à Morin [16] pp 47-48. Voir aussi [[12] , Th. 2] .

PROPOSITION 1.4. *Soit $F : C' \longrightarrow B'$ un morphisme propre et plat de variétés lisses. On suppose qu'en dehors d'une sous variété lisse D de codimension 1 dans B', F est un fibré en coniques et que sur D les fibres sont des coniques dégénérées réduites (i.e., réunion de deux courbes rationnelles lisses se coupant en un point). Soit $B = B' - D$, $C = F^{-1}(B)$ et $\pi = F|C$. Si $F^{-1}(D)$ est irréductible alors le fibré en coniques $F : C \longrightarrow B$ n'est pas banal.*

DEMONSTRATION. Remarquons d'abord que F définit un revêtement $\tilde{D}$ à deux feuillets de D (on peut définir $\tilde{D}$ comme le schéma de Hilbert relatif des droites contenues dans les fibres sur D) ; de plus il existe un morphisme naturel $\eta : F^{-1}(D) - \mathcal{S} \longrightarrow \tilde{D}$, ou $\mathcal{S}$ est l'ensemble des points singuliers des fibres.

Supposons que $C \longrightarrow B$ est banal. Soit $\tau : U' \longrightarrow C$ une section sur un ouvert non-vide U de B. Comme F est propre, τ s'étend comme une section $\sigma : U \longrightarrow C'$ où U est un ouvert de B' avec codim $(B'-U) \geq 2$. Si on pose $W = U \cap D$, alors W est un ouvert non-vide de D. Si $w \in W$, $\sigma(w)$ ne peut pasêtre le point singulier de $F^{-1}(w)$ puisque F est lisse en $\sigma(w)$. En composant $\sigma|W$ avec η, on aura une section de $\tilde{D} \longrightarrow D$ sur W. Puisque D est lisse, cette section s'étend en une section de $\tilde{D} \longrightarrow D$ sur D. Ainsi le revêtement $\tilde{D} \longrightarrow D$ se scinde, ce qui démontre que $F^{-1}(D)$ est réductible.

EXEMPLE 1.5. Soit B le schéma de Hilbert des coniques lisses dans $\mathbb{P}^2$ et $C \longrightarrow B$ le fibré en coniques tautologique. Alors $C \longrightarrow B$ n'est pas banal. En effet, soit $F : C' \longrightarrow B'$ la famille tautologique des coniques réduites dans $\mathbb{P}^2$ et $D \subset B'$ le diviseur paramétrant les coniques singulières. Alors $F^{-1}(D)$ est irréductible ; par exemple, parce que PGL(3) opère transitivement sur le revêtement $\tilde{D}$.

2. FIBRÉ EN GRASSMANNIENNES ASSOCIÉ A UN FIBRÉ EN CONIQUES.

Soit $\pi : C \longrightarrow B$ un fibré en coniques. Soit k un entier ≥ 1. On désigne par $C^{(k)}$ le schéma de Hilbert $\mathrm{Hilb}^k\, C/B$ des diviseurs positifs de degré k sur les fibres de $C \longrightarrow B$. Alors $C^{(k)} \longrightarrow B$ est un fibré projectif (de fibre $\mathbb{P}^k$).

REMARQUE 2.1. Si k est pair, le fibré projectif $C^{(k)} \longrightarrow B$ est banal. Il est associé à l'image directe du faisceau $\overset{k/2}{\otimes} T_\pi$, où T_π est le fibré tangent relatif.

On désigne par GC(d,k) la grassmannienne relative des sous-espaces projectifs de dimension d des fibres de $C^{(k)}$ $(0 \leq d \leq k)$. On note M(d,k) l'ouvert de GC(d,k) formé des systèmes linéaires de dimension d sans point de base.

PROPOSITION 2.2. *Supposons que k est pair ou que k et d sont tous les deux impairs. Soit G(d,k) la grassmannienne des sous espaces de dimension d de l'espace projectif $\mathbb{P}^k$. Alors il existe un ouvert non-vide U de B, un ouvert affine non-vide A de G(d,k) et un plongement ouvert de U X A dans GC(d,k) au dessus de B.*

En particulier si B est rationnelle, GC(1,k) et M(1,k) sont rationnelles.

On va démontrer d'abord des lemmes.

LEMME 2.3. *Tout fibré de fibre l'espace affine A^n de groupe structural le groupe affine est localement trivial dans la topologie de Zariski.*

DÉMONSTRATION. On va démontrer que chaque fibré principal de groupe structural le groupe affine est localement trivial dans la topologie de Zariski. D'après [17, Théorème 2, § 4.3], il suffit de démontrer que le groupe affine est spécial i.e., qu'il existe un plongement du groupe affine H dans GL(N) comme sous groupe fermé tel que GL(N) $\longrightarrow$ GL(N)/H admette une section rationnelle. Mais le groupe affine est le sous groupe de GL(n+1) formé par les matrices de la forme

$$\begin{pmatrix} T & * \\ 0 & 1 \end{pmatrix} \quad \text{où } T \in GL(n)$$

Le sous-espace de GL(n+1) des matrices de la forme

$$\begin{pmatrix} I & 0 \\ * & \mu \end{pmatrix}$$

où I est la n x n matrice identique et $\mu \neq 0$, donne une section rationnelle.

LEMME 2.4. *Soit V un sous-espace projectif de dimension (k-d-1) dans $\mathbb{P}^k$, et $G^V(d,k)$ l'ouvert de la grassmannienne G(d,k) des sous espaces projectifs de dimension d disjoints de V. Alors $G^V(d,k)$ admet une structure d'espace affine pour laquelle l'action sur $G^V(d,k)$ du sous groupe* *de PGL(k+1) qui stabilise V est affine.*

DÉMONSTRATION. Soit $\mathbb{P}^k = \mathbb{P}(F)$ où F est un espace vectoriel de dimension k+1 et $\underline{V}$ le sous espace vectoriel de F correspondant à V. Alors $G^V(d,k)$ s'identifie à l'espace des scindages de la suite exacte d'espaces vectoriels

$$0 \longrightarrow \underline{V} \longrightarrow F \longrightarrow F/\underline{V} \longrightarrow 0,$$

qui est un espace affine (sous l'action de $\mathrm{Hom}(F/\underline{V},\underline{V})$. Soit $\underline{W}$ un supplémentaire de $\underline{V}$. Alors $G^V(d,k)$ s'identifie à $\mathrm{Hom}(\underline{W},\underline{V})$. Les éléments de GL(k+1) qui stabilisent V s'écrivent

$$\begin{pmatrix} f & g \\ o & h \end{pmatrix}$$

On vérifie que pour x dans $\mathrm{Hom}(\underline{W},\underline{V})$ $\begin{pmatrix} f & g \\ o & h \end{pmatrix}(x)$ est égal à $fxh^{-1} + gh^{-1}$, ce qui est bien affine en x.

LEMME 2.5. *Supposons que ℓ et k ont même parité. Alors $GC(\ell,k)$ admet une section rationnelle.*

DÉMONSTRATION. Comme $(k-\ell)$ est pair, $C^{(k-\ell)} \longrightarrow B$ admet une section rationnelle σ (Remarque 2.1). On a un morphisme naturel $\varphi : C^{(k-\ell)} \times_B C^{(\ell)} \longrightarrow C^{(k)}$ ("addition des diviseurs"). Soit $(pr_1,\varphi) : C^{(k-\ell)} \times_B C^{\ell} \longrightarrow C^{(k-\ell)} \times C^{(k)}$ le morphisme associé. C'est un plongement dont l'image est propre et plate sur $C^{(k-\ell)}$. Il définit un B-morphisme $\psi : C^{(k-\ell)} \longrightarrow GC(\ell,k)$. Alors $\tau = \psi o \sigma$ est la section voulue.

DÉMONSTRATION DE LA PROPOSITION 2.2.

Si k est pair le résultat est évident, car GC(d,k) est localement trivial (dans la topologie de Zariski), $C^{(k)} \longrightarrow B$ étant localement trivial (Remarque 2.1). On va supposer k et d impairs. D'après le lemme 2.5, GC(k-d-1,k) admet une section σ sur un ouvert non-vide U' de B. Soit M le sous-groupe de PGL(k+1) qui stabilise un (k-d-1) plan V dans $\mathbb{P}^k$. Si P est le fibré principal de groupe PGL(k+1) associé au fibré projectif $C^{(k)} \longrightarrow B$, la section σ permet de définir une réduction du groupe structural de $P|U'$ à M. Soit R le fibré principal sur U' avec groupe M ainsi obtenu. Avec la notation du Lemme 2.4, soit E le fibré associé à R avec fibre $G^V(d,k)$, à partir de l'action de M sur $G^V(d,k)$. Alors E est un fibré affine (Lemme 2.4) et il y a un plongement ouvert de E dans GC(d,k) au dessus de U'. D'après le Lemme 2.3, le fibré affine E est localement trivial sur U' (dans la topologie de Zariski), ce qui démontre la proposition.

PROPOSITION 2.6. *Soit $p : M(1,k) \longrightarrow B$ la projection canonique. Si le fibré projectif $C \longrightarrow B$ n'est pas banal, alors le fibré projectif $p^*(C) \longrightarrow M(1,k)$, image réciproque de $C \longrightarrow B$ par p, n'est pas banal non plus.*

DÉMONSTRATION. Avec la notation de la proposition 2.2, soit $\Omega \subset GC(1,k)$ l'image de U x A. Comme GC(1,k) est irréductible il existe $\xi \in \Omega \cap M(1,k)$. Soit σ une section rationnelle de $\Omega \longrightarrow B$ passant par ξ (il en existe puisque $\Omega \approx U \times A$). Alors pour un voisinage W de $p(\xi)$ dans B, on aura $\sigma(W) \in M(1,k)$. Si $p^*(C) \longrightarrow M(1,k)$ était banal, C serait banal sur W et donc sur B (Proposition 1.1).

3. LE REVÊTEMENT A DEUX FEUILLES DE L'ESPACE DES QUADRIQUES LISSES DANS $\mathbb{P}^3$ ET FIBRÉS ASSOCIÉS.

Soit $\mathcal{H}$ le schéma de Hilbert des quadriques lisses dans $\mathbb{P}^3$. Soient $\tilde{\mathcal{H}}$ le revêtement à deux feuilles de $\mathcal{H}$ correspondant au choix d'un système de génératrices sur les quadriques et $C \longrightarrow \tilde{\mathcal{H}}$ le fibré projectif dont la fibre en $\xi \in \tilde{\mathcal{H}}$ est la droite projective des génératrices correspondant à ξ [3,§ 2.7] .

On a le diagramme

$$\begin{array}{ccccc} C & \xleftarrow{\;g\;} & \tilde{Q} & \hookrightarrow & \tilde{\mathcal{H}} \times \mathbb{P}^3 \\ & {\scriptstyle p}\searrow & & \swarrow & \\ & & \tilde{\mathcal{H}} & & \end{array}$$

où $\tilde{Q}$ est l'image réciproque de la quadrique universelle $Q \longrightarrow \mathcal{H}$.

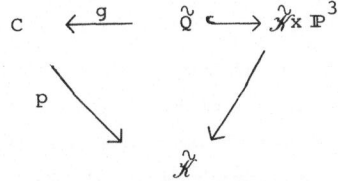

DEFINITION 3.1. Soit $C \longrightarrow \tilde{\mathcal{X}}$ le fibré projectif défini ci-dessus. On désigne par M_k l'ouvert de $GC(1,k)$ formé par les systèmes linéaires de dimension 1 sans points de base, où $GC(1,k)$ est la grassmannienne relative des sous-espaces projectifs de dimension 1 des fibres de $C^{(k)}$. [$C^{(k)}$ est le fibré projectif $\mathrm{Hilb}^k(C/\tilde{\mathcal{X}})$, voir § 2.] . On va noter p la projection canonique $M_k \longrightarrow \tilde{\mathcal{X}}$.

REMARQUE 3.2. Soit $\pi : Q_S \longrightarrow S$, $Q_S \subset S \times \mathbb{P}^3$, une famille de quadriques lisses sur une variété S. Alors

1) - Si F est un fibré de rang un et de bidegré $\{0,k\}$, $k \geq 1$, sur Q_S, (Q_S,F) définit un morphisme de S dans $\tilde{\mathcal{X}}$.

2) - Si D est un fibré de rang 2 sur S et $\delta : \pi^*(D) \longrightarrow F$ est un morphisme surjectif, δ définit un morphisme, noté $[\delta]$, de S dans M_k.

3) - Soient F', D' et $\delta' : \pi^*(D') \longrightarrow F'$ vérifiant les mêmes hypothèses avec F et F' isomorphes. Si $[\delta] = [\delta']$, alors δ et δ' sont isomorphes.

LEMME 3.3. *Soit S une variété et $\pi : Q_S \longrightarrow S$, $Q_S \subset S \times \mathbb{P}^3$, une famille lisse de quadriques sur S. Soit F un fibré en droites de bidegré $\{0,k\}$, $k \geq 1$, sur Q_S et $\varphi : S \longrightarrow M_k$ un morphisme tel que $p \circ \varphi : S \longrightarrow \tilde{\mathcal{X}}$ soit le morphisme associé à (Q_S, F). Alors il existe un sous fibré (unique) D de rank 2 de $\pi_* F$ et un morphisme $\delta : \pi^*(D) \longrightarrow F$ tel que le morphisme $[\delta]$ associé vérifie $[\delta] = \varphi$.*

DEMONSTRATION. On considère le diagramme commutatif

$$\begin{array}{ccccc} C & \longleftarrow & C_S & \xleftarrow{\;g_S\;} & Q_S \\ & {\scriptstyle p}\searrow & & {\scriptstyle p_S}\searrow \quad \swarrow{\scriptstyle \pi_S} & \\ & & \tilde{\mathcal{X}} & \xleftarrow[\;p\circ\varphi\;]{} & S \end{array}$$

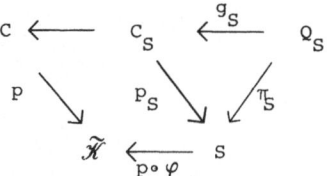

Alors $(g_S)_* F$ est un fibré en droites de degré relatif k sur C_S. Or $(p_S)_*((g_S)_* F)$ est un fibré vectoriel E de rang k+1 sur S et le fibré projectif $\mathbb{P}(E)$ est canoniquement isomorphe au fibré projectif $C_S^{(k)}$. Cet isomorphisme induit un isomorphisme τ de $GC_S(1,k)$ (formé à partir de C_S) sur la variété grassmannienne relative $GE(2)$ des sous-espaces vectoriels de dimension 2 des fibres de E. Soit W le (sous) fibré tautologique de rang 2 sur GE(2) et $\tilde{\varphi}$ la section de $GC_S(1,k)$ induite par φ . Alors $D=(\tau \circ \tilde{\varphi})^*W$ est le fibré cherché. ($\pi^*D \longrightarrow L$ est surjectif car pour $s \in S$, $\varphi(s)$ est un système linéaire sans point de base).

THEOREME 3.4.

I) *Le fibré projectif* $C \longrightarrow \tilde{\mathcal{K}}$ *n'est pas banal.*

II) $\tilde{\mathcal{K}}$ *est une variété rationnelle.*

III) M_k *(définition 3.1) est une variété rationnelle.*

IV) *Le fibré projectif* $p^*C \longrightarrow M_k$ *n'est pas banal (ici,* $p : M_k \longrightarrow \tilde{\mathcal{K}}$ *est la projection naturelle).*

DEMONSTRATION DE 3.4. I). Le résultat découle du lemme suivant en utilisant le fait qu'il existe des fibrés en coniques qui ne sont banals sur aucun ouvert non-vide de la base (Exemple 1.5 et Proposition 1.1).

LEMME 3.5. *Soit* $P \longrightarrow S$ *un fibré en coniques. Pour chaque* $s_0 \in S$ *il existe un voisinage* U *de* s_0 *et un morphisme* $\tilde{\varphi}: U \longrightarrow \tilde{\mathcal{K}}$ *tel que* $P|U$ *soit isomorphe à l'image réciproque de* $C \longrightarrow \tilde{\mathcal{K}}$ *par* φ.

DEMONSTRATION. Il suffit de démontrer qu'on peut choisir un fibré vectoriel W de rang 4 sur S et une famille de quadriques lisses $Q \longrightarrow S$, $Q \subset \mathbb{P}(W)$, munie d'un système de génératrices, tel que le fibré P soit isomorphe au fibré en coniques défini par les génératrices choisies. Car, en trivialisant W au voisinage de s_0, on aura l'application φ cherchée, vu la propriété universelle de $\tilde{\mathcal{K}}$.

Soit T le fibré principal avec groupe structural PGL(2) associé à P. On considère l'action diagonale de PGL(2) sur la quadrique $\mathbb{P}^1 \times \mathbb{P}^{1\vee} \subset \mathbb{P}(V \otimes V^\vee)$ ou $\mathbb{P}^1 = \mathbb{P}(V)$, V étant un espace vectoriel de dimension 2. L'action de PGL(2) sur l'espace des génératrices paramétré par $\mathbb{P}^1$ correspond à l'action naturelle de $\mathbb{P}$GL(2) sur $\mathbb{P}^1$. Soit W le fibré vectoriel associé à T par l'action linéaire de PGL(2) sur $V \otimes V^\vee$. Si $Q \longrightarrow S$, est le fibré en quadriques associé à T par l'action de PGL(2) sur $\mathbb{P}^1 \times \mathbb{P}^{1\vee}$, on a $Q \subset \mathbb{P}(W)$ et Q vient avec le choix d'un système de génératrices sur chaque fibre. On vérifie que Q a la propriété voulue.

REMARQUE. W est en fait l'algèbre d'Azumaya associée au fibré projectif P.

Pour la clarté de l'exposé on va démontrer d'abord un résultat analogue à 3.4.II, pour le revêtement à deux feuilles de l'espace des formes quadratiques non-dégénérées, et puis on va indiquer les modifications nécessaires dans le cas de l'espace des quadriques non-singulières.

PROPOSITION 3.6. *Soit $\mathcal{B}$ l'espace des formes quadratiques non-dégénérées sur $V = k^4$ (k un corps algébriquement clos). Soit $\tilde{\mathcal{B}}$ le revêtement à deux feuilles associé au choix d'un système de sous espaces totalement singuliers maximaux. Alors $\tilde{\mathcal{B}}$ est une variété rationnelle.*

Avant de démontrer cette proposition nous rappelons ci-dessous la description explicite de la $k(\mathcal{B})$-algèbre $k(\tilde{\mathcal{B}})$ puis nous donnons une condition suffisante pour qu'une extension d'ordre deux d'une extension transcendante pure de k soit une extension transcendante pure de k. La démonstration de 3.6 consistera à vérifier cette condition, ce qui est immédiat si car $k \neq 2$.

RAPPEL 3.7.

1°) - (cf [3] Prop.2.8). Soit Q une forme quadratique non dégénérée sur K^4, C(Q) son algèbre de Clifford. Le centre Z de la partie paire de C(Q) s'identifie à l'anneau des fonctions rationnelles du K-schéma des systèmes de sous-espaces totalement singuliers maximaux de Q.

2°) - (cf [2] § 9 n° 4 Remarque 2 et Exercice 9) : Si K est un corps, Q une forme quadratique non dégénérée sur K^4, C(Q) son algèbre de Clifford et Z le centre de la partie paire de C(Q), alors:

A) Si car $K \neq 2$, il existe y dans Z tel que $\{1,y\}$ engendre Z sur K et que y^2 soit le discriminant de Q.

B) Si car K = 2, soit (e_1,e_2,e_3,e_4) une base symplectique pour la forme bilinéaire (alternée!) associée à Q ; alors l'élément $y=e_1e_2+e_3e_4$ de C(Q) est dans Z, vérifie $y^2+y=Q(e_1)Q(e_2)+Q(e_3)Q(e_4)$, et est tel que $\{1,y\}$ engendre Z sur K.

LEMME 3.8. *Soit k un corps et $K = k(x_1,\ldots,x_N)$ une extension transcendante pure de k avec $\{x_1,\ldots,x_N\}$ algébriquement libre sur k. Soit P un polynôme de $x_1,\ldots,x_N$ de la forme $P = x_1 T+R$ où T et R sont des polynômes de $x_2,\ldots,x_N$ et $T\neq 0$. Soit Z une algèbre commutative sur K admettant une base $\{1,y\}$ où y vérifie l'équation $y^2+\lambda y = P$, $\lambda\in k$. Alors Z est un corps et c'est une extension transcendante pure de k.*

DÉMONSTRATION. On vérifie que P n'est pas de la forme $\xi^2+\lambda\xi = P$ avec $\xi \in K$, d'où Z est un corps. Comme $x_1 = \frac{P-R}{T}$ on a $K = k(P,x_2,\ldots,x_N)$ et $\{P,x_2,\ldots,x_N\}$ est algébriquement libre sur k. Alors $Z = k(y, x_2,\ldots,x_N)$ avec $\{y,x_2,\ldots,x_N\}$ algébriquement libre sur k.

DEMONSTRATION DE LA PROPOSITION 3.6.*

Le cas où car $k \neq 2$ résulte immédiatement de 3.7 et 3.8. On suppose dans la suite que car $k = 2$.

Soit $(v_1,\ldots,v_4)$ la base canonique de $V = k^4$. Si L est un surcorps de k on va noter aussi $(v_1,\ldots,v_4)$ la base $(v_1\otimes 1_L,\ldots,v_4\otimes 1_L)$ de $V \otimes_k L$.

Soit K le corps des fonctions rationnelles sur $\mathcal{B}$. Si $\{x_{ij}\}_{i\leq j}$, $1 \leq i \leq 4$, $1 \leq j \leq 4$ sont des indéterminées sur k on a $K = k(x_{ij})$ et la forme quadratique universelle Q sur K est donnée par :

$$Q\left(\sum_{i=1}^{4} \xi_i v_i\right) = \sum_{i=1}^{4} x_{ii}\,\xi_i^2 + \sum_{i<j} x_{ij}\,\xi_i\,\xi_j,\quad \xi_i \in K.$$

Considérons le sous-corps K' de K engendré par les x_{ij} avec $i < j$ [K' est le corps de fonctions de l'espace des formes binéaires alternées sur k^4]. La forme alternée universelle A (définie sur K') est donnée par la matrice (a_{ij}), avec $a_{ij} = x_{ij}$ pour $i < j$, $a_{ji} = x_{ij}$ pour $j > i$ et $a_{ii} = 0$. De plus A est la forme bilinéaire alternée associée à Q. Choisissons une base symplectique $(e_1,\ldots,e_4)$ pour A dans $V \otimes_k K'$.

* Nous remercions M.V. Nori et A. Ramanathan de l'aide qu'ils nous ont apportée pour cette démonstration.

En utilisant 3.7.1, 3.7.2 B et 3.8, on peut terminer la démonstration de la proposition si on montre que $\{Q(e_1), Q(e_2), Q(e_3), Q(e_4)\}$ forme une base transcendante pure de K sur K' (car on aura alors $K = k(x_{ij}, Q(e_k))$, $i < j$, $1 \leq k \leq 4$). Pour le montrer il suffit de prouver que le sous-corps de K engendré sur K' par les $Q(e_i)$ est identique à K, puisque $K = K'(x_{11},\ldots,x_{44})$. Ecrivons

$$v_i = \sum_j \xi_{ij}\, e_j,\ \xi_{ij} \in K'.$$

On a alors

$$x_{ii} = Q(v_i) = \sum_{j<k} \xi_{ij}\, \xi_{ik}\, A(e_j,e_k) + \sum_j \xi_{ij}^2\, Q(e_j).$$

Comme $\xi_{ij} \in K'$ et $A(e_j,e_k) \in k$, on voit que $K'(Q(e_i)) = K'(x_{ii})$. Ceci complète la démonstration de 3.6.

DÉMONSTRATION DE 3.4.II.

La méthode de la démonstration de 3.4.II est analogue à celle de 3.8. Soit K le corps des fonctions rationnelles sur $\mathscr{K}$. L'idée est de choisir une section rationnelle de $\mathscr{B} \longrightarrow \mathscr{K}$ et de travailler avec la forme quadratique induite de Q sur K par cette section. On prendra la section donnée par les formes quadratiques q sur k^4 avec $q(v_1)=1$ i.e. définie par $x_{11} = 1$.

Avec la notation de la démonstration 3.6, le corps K des fonctions rationnelles sur $\mathscr{K}$ est isomorphe à $k(x_{22},x_{33},x_{44},x_{ij})$, $i < j$. Considérons la forme quadratique Q sur K :

$$Q\Big(\sum_{i=1}^{4} \xi_i\, v_i\Big) = \xi_1^2 + \sum_{j=2}^{4} x_{jj}\, \xi_j^2 + \sum_{i<j} x_{ij}\, \xi_i\, \xi_j .$$

Quand car $k \neq 2$, en utilisant 3.7.1, 3.7.2.A et 3.8, on voit, comme dans la démonstration de 3.6, que $\tilde{\mathscr{K}}$ est rationnelle. Supposons donc car $k = 2$. Dans ce cas, K contient K' et la forme bilinéaire alternée associée à Q est A. Choisissons une base symplectique $(e_1,\ldots,e_4)$ dans $V \otimes_k K'$ avec $e_1 = v_1$. On voit comme dans la démonstration de 3.6 que $\{Q(e_2), Q(e_3), Q(e_4)\}$ est une base pure de K sur K'.
On termine en utilisant 3.8.

REMARQUE 3.9. On peut démontrer 3.4. II, en montrant, au moins en caractéristique nulle, que $\tilde{\mathscr{K}}$ est isomorphe à l'ouvert de la grassmannienne des plans dans $\mathbb{P}^5$, constitué par les plans qui coupent la quadrique de Klein suivant une conique lisse. Mais la démonstration donnée plus haut s'étend facilement au cas des quadriques de $\mathbb{P}^{2m+1}$, $m \geq 1$.

DÉMONSTRATION DE 3.4., III. C'est une conséquence immédiate de 3.4.II et de la proposition 2.2.

DÉMONSTRATION DE 3.4. IV. 3.4, IV résulte de 3.4.I et de la proposition 2.6.

4. FIBRÉS DE 't HOOFT SPÉCIAUX ('tHS).

DEFINITION 4.1. Un fibré vectoriel E sur $\mathbb{P}^3$ est un fibré 'tHS (fibré de 't Hooft spécial) s'il existe une quadrique non-singulière Q_E et un fibré $\mathscr{F}_E$ de rang un sur Q_E tel que E soit extension du prolongement $\tilde{\mathscr{F}}_E$ de $\mathscr{F}_E$ à $\mathbb{P}^3$ par un fibré trivial de rang 2.

REMARQUE 4.2. Les fibrés de 't Hooft sont les fibrés admettant une section dont le schéma des zéros est une réunion disjointes de droites (réduites). On peut montrer en caractéristique nulle que les fibrés 'tHS sont les fibrés de 't Hooft correspondant aux réunions disjointes de droites contenues dans une même quadrique non-singulière.

PROPOSITION 4.3. *Soit E un fibré 'tHS avec* $c_2 \geq 3$. *Alors* Q_E *et* $\mathscr{F}_E$ *sont uniquement déterminés (à isomorphisme près pour* $\mathscr{F}_E$*). De plus,* $H^0(\mathbb{P}^3,E)$ *est de dimension deux et* $\tilde{\mathscr{F}}_E$ *est le conoyau de l'injection naturelle*

$$j_E = H^0(\mathbb{P}^3,E) \otimes \mathcal{O} \longrightarrow E.$$

Le bidegré de $\mathscr{F}_E$ *est* $\{2, 2-c_2\}$.

DÉMONSTRATION. Soit $\{a,b\}$ le bidegré de $\mathscr{F}_E$. Le polynôme de Hilbert de $\tilde{\mathscr{F}}_E$ est donc

$$\chi(t) = (t + a + 1)\ (t + b + 1)$$

et la formule de Riemann-Roch donne

$c_1(\tilde{\mathcal{F}}_E)=2$, $c_2(\tilde{\mathcal{F}}_E) = 4 - a - b$, $c_3(\tilde{\mathcal{F}}_E) = 2(a-2)(b-2)$.

Comme ce sont aussi les classes de Chern de E on doit avoir $c_3(\tilde{\mathcal{F}}_E)=0$ donc par exemple a=2 et la condition $c_2 \geq 3$ devient $b < 0$. On voit alors que $\tilde{\mathcal{F}}_E$ n'a pas de section globale, ce qui prouve que le morphisme donné $\mathcal{O}\oplus\mathcal{O} \longrightarrow E$ s'identifie au morphisme j_E de l'énoncé : $\tilde{\mathcal{F}}_E$ s'identifie donc au conoyau de j et Q_E est le support de ce conoyau.

PROPOSITION 4.4. *Soit S une variété, π la projection $S \times \mathbb{P}^3 \longrightarrow S$ et $\mathcal{E}$ un fibré sur $S\times\mathbb{P}^3$ tel que pour tout s dans S, $\mathcal{E}(s)$ soit 'tHS avec $c_2 \geq 3$. On note $j_{\mathcal{E}}$ le morphisme naturel de $\pi^* \pi_* \mathcal{E}$ dans $\mathcal{E}$. Alors les quadriques $Q_{\mathcal{E}(s)}$ s'organisent en une famille plate notée $Q_{\mathcal{E}}$ et le conoyau de $j_{\mathcal{E}}$ est le prolongement par zéro d'un fibré de bidegré $\{2, 2-c_2\}$ sur $Q_{\mathcal{E}}$.*

DEMONSTRATION. Nous allons faire apparaitre $Q_{\mathcal{E}}$ comme support du conoyau de det $j_{\mathcal{E}}$.

D'après le théorème de changement de base, $\pi_*(\mathcal{E})$ est un fibré de rang deux ainsi que $\pi^* \pi_* \mathcal{E}$ et $\pi^* \pi_* \mathcal{E}(s)$ s'identifie à $H^0(\mathbb{P}^3, \mathcal{E}(s)) \otimes \mathcal{O}$, cependant que $j_{\mathcal{E}}(s)$ s'identifie au morphisme d'évaluation :

$$j_{\mathcal{E}}(s) = j_{\mathcal{E}(s)} : H^0(\mathbb{P}^3, \mathcal{E}(s)) \otimes \mathcal{O} \longrightarrow \mathcal{E}(s).$$

Comme, pour les fibrés, l'élévation à une puissance extérieure et la restriction commutent, ces identifications s'étendent au morphisme

$$\det j_{\mathcal{E}} : \Lambda^2 \pi^* \pi_* \mathcal{E} \longrightarrow \Lambda^2 \mathcal{E}$$

de sorte que det $j_{\mathcal{E}}(s)$ s'identifie à

$$\det j_{\mathcal{E}(s)} : \Lambda^2 H^0(\mathbb{P}^3, \mathcal{E}(s)) \otimes \mathcal{O} \longrightarrow \Lambda^2 \mathcal{E}(s).$$

Or ce dernier morphisme est injectif pour tout s. Par suite, si H désigne le conoyau de det $j_{\mathcal{E}}$, on voit que le polynôme de Hilbert de H(s) est indépendant de s. De ce fait H est S-plat [5 , Th III, 9.9 et sa preuve] , et $H \otimes \Lambda^2\mathcal{E}$ est le faisceau structural de la famille $Q_{\mathcal{E}}$ cherchée. Nous allons maintenant voir que le conoyau coker $j_{\mathcal{E}}$ est le prolongement par zéro de sa restriction à $Q_{\mathcal{E}}$. Cela résulte du

LEMME 4.5. *Soit A et B deux fibrés de même rang sur une variété X et j : A ⟶ B un morphisme injectif de faisceaux. Alors l'idéal défini par det j annule coker j, autrement dit : si Q est le sous-schéma défini par cet idéal, coker j s'identifie au prolongement par zero de sa restriction à Q.*

DÉMONSTRATION. L'énoncé étant local, on peut supposer que A et B sont triviaux. Les formules de Cramer donnent alors un morphisme $\mathrm{Cof}_j : B \longrightarrow A$ vérifiant $j \circ \mathrm{Cof}_j = \det j.\ \mathrm{Id}_B$ ce qui prouve que $\det_j(B)$ est dans l'image de A.

Pour terminer la démonstration de la Proposition 4.4., il nous suffit maintenant de prouver que la restriction à $Q_{\mathcal{E}}$ de $j_{\mathcal{E}}$ a pour conoyau un fibré de rang un (il aura le bon bidegré d'après la Proposition 4.3).

Comme $Q_{\mathcal{E}}$ est lisse, det $j_{\mathcal{E}}$ s'annule à l'ordre un sur $Q_{\mathcal{E}}$. Par suite $j_{\mathcal{E}}(x)$ est de rang un en tout point x de $Q_{\mathcal{E}}$. Pour conclure, il nous suffit d'appliquer le résultat général suivant : soient A et B deux fibres de rangs r_A et r_B sur la variété X et j : A ⟶ B un morphisme de fibrés de rang constant r. ALors ker j, Im j, coker j sont des fibrés de rangs $r_A - r$ et $r_B - r$, dont la construction commute avec les changements de base.

Dans la suite de ce paragraphe nous fixons $c_2 \geq 3$ et nous étudions le foncteur 'tHS(c_2), ou plus simplement 'tHS, sur la catégorie des variétés, qui à S associe l'ensemble des classes d'isomorphisme de fibrés sur $S \times \mathbb{P}^3$ vérifiant l'hypothèse de la Proposition 4.4. Notons que cette proposition nous permettrait d'étendre le foncteur 'tHS à la catégorie des schémas, mais cela n'est pas utile ici.

Nous voulons montrer que la variété $M = M_{c_2}$ considérée en 3.1 est un module grossier pour 'tHS. Pour cela il nous faut le

LEMME 4.6. *Si F est le prolongement par zéro d'un fibré de rang un et bidegré {a,b} sur une famille plate de quadriques non-singulières Q sur S alors $\mathcal{E}xt^1(F,\mathcal{O})$ est le prolongement par zèro d'un fibré de rang un sur Q de bidegré {2-a, 2-b} .*

DÉMONSTRATION. L'affirmation étant locale dans S on peut supposer que Q admet une équation

$$0 \longrightarrow \mathcal{O}(-2) \xrightarrow{q} \mathcal{O} \longrightarrow \mathcal{O}_Q \longrightarrow 0.$$

D'où

$$0 \longrightarrow \mathcal{O}(a-2) \xrightarrow{q} \mathcal{O}(a) \longrightarrow \mathcal{O}_Q(a) \longrightarrow 0$$

et en dualisant

$$\mathcal{O}(-a) \xrightarrow{q} \mathcal{O}(2-a) \longrightarrow \mathcal{E}xt^1(\mathcal{O}_Q(a),\mathcal{O}) \longrightarrow 0$$

Ceci prouve que $\mathcal{E}xt^1(\mathcal{O}_Q(a),\mathcal{O})$ est le prolongement par zéro d'un fibré de rang un sur Q de bidegré $\{2-a,2-a\}$. Il en résulte que $\mathcal{E}xt^1(F,\mathcal{O})$ est le prolongement par zéro d'un fibré de rang un sur Q. Le bidegré $\{2-a, 2-b\}$ de ce fibré se calcule en considérant des ouverts (non-affines) de $S \times \mathbb{P}^3$ où F est isomorphe à $\mathcal{O}_Q(a)$ où à $\mathcal{O}_Q(b)$.

DÉFINITION 4.7. *Soit $\mathcal{E}$ comme en 4.4 et $\delta : (\pi^* \pi_* \mathcal{E})^\vee \longrightarrow \mathcal{E}xt^1(\tilde{\mathcal{F}}_{\mathcal{E}},\mathcal{O})$ le conoyau de $j^\vee$. D'après le lemme 4.6 et la remarque 3.2, $\boxed{\delta}$ définit un morphisme de foncteurs, noté mod, de 'tHS dans le foncteur $\bar{M}$ représenté par M, où $M = M_{c_2}$ (Définition 3.1).*

L'image du morphisme mod est caractérisé par l'énoncé suivant.

PROPOSITION 4.8. Pour qu'un morphisme $\varphi : S \longrightarrow M$ soit dans l'image de mod (i.e. provienne d'une famille de fibrés 'tHS) il faut et il suffit qu'il existe sur le fibré $\varphi^* p^* C$ un fibré de rang un et de degré relatif c_2, où $p : M \longrightarrow \tilde{\mathcal{K}}$ est la projection canonique.

DEMONSTRATION. La condition est nécessaire ; en effet si $\mathcal{E}$ est un fibré 'tHS sur $S \times \mathbb{P}^3$ tel que mod $\mathcal{E} = \varphi$, alors d'après le lemme 4.6 $\mathcal{E}xt^1(\tilde{\mathcal{F}}_{\mathcal{E}},\mathcal{O})$ provient d'un fibré de rang un et de degré c_2 sur $\varphi^* p^* C$.

La condition est suffisante : Soit L un fibré de rang un et degré relatif c_2 sur $C_S = \varphi^* p^* C$. On considère le diagramme déduit du diagramme universel (§ 3),

$$\begin{array}{ccccc} C_S & \xleftarrow{g_S} & Q_S & \xrightarrow{i_S} & S \times \mathbb{P}^3 \\ & {\scriptstyle p_S}\searrow & \downarrow{\scriptstyle q_S} & \swarrow{\scriptstyle \pi_S} & \\ & & S & & \end{array}$$

D'après le lemme 3.3 (et sa preuve) φ définit un sous fibré de rang 2, noté D, de $p_{S_*} L$ auquel correspond un morphisme

$$\delta' : p_S^* D \longrightarrow L$$

vérifiant $[\delta'] = \varphi$.

Ce morphisme induit un morphisme surjectif

$$\delta : \pi_S^* D \longrightarrow i_{S_*} g_S^* L.$$

On pose $\mathscr{E} = (\ker \delta)^{\vee}$ et on montre que $\mathscr{E}$ est un fibré 'tHS vérifiant mod $\mathscr{E} = \varphi$. D'abord Ker δ est un fibré parce que $i_{S_*} g_S^* L$ est de dimension homologique un. En dualisant la suite

$$0 \longrightarrow \mathscr{E}^{\vee} \xrightarrow{\alpha} \pi_S^* D \longrightarrow i_{S_*} g_S^* L \longrightarrow 0$$

on obtient

$$0 \longrightarrow \pi_S^* D^{\vee} \xrightarrow{\alpha^{\vee}} \mathscr{E} \longrightarrow \mathscr{E}xt^1(i_{S_*} g_S^* L, \mathcal{O}) \longrightarrow 0$$

ce qui prouve que $\mathscr{E}$ est un fibré 'tHS. On a alors par définition mod $\mathscr{E} = [\text{coker } \alpha^{\vee\vee}]$ et comme $\alpha^{\vee\vee} = \alpha$, mod $\mathscr{E} = [\delta]$. Or $[\delta] = [\delta']$.

Pour montrer que M est un module grossier nous utiliserons le

<u>LEMME 4.9.</u> Soit F un foncteur sur la catégorie des variétés et mod : $F \longrightarrow \hat{M}$ un morphisme de F vers le foncteur représenté par M. On suppose

1) - pour toute variété S, si α et β sont deux éléments de F(S) vérifiant mod α = mod β , alors α et β sont localement égaux dans S.

2) - il existe T et u dans F(T) tel que mod u soit propre et lisse.

Alors mod fait de M un module grossier.

DEMONSTRATION. Observons que $T \times_M T$ est une variété et que les deux éléments u_1 et u_2 de $F(T\times_M T)$ induits par u sont localement égaux, d'après la première hypothèse. Par suite, si $n: F \longrightarrow \hat{N}$ est un autre morphisme de foncteurs, alors $n(u_1)$ et $n(u_2)$ sont égaux. Comme $T \longrightarrow M$ est un morphisme de descente, cela prouve que $n(u) : T \longrightarrow N$ se factorise à travers M. Soit $f : M \longrightarrow N$ le morphisme correspondant. Soit maintenant $v \in F(S)$ et montrons $n(v) = f \circ \text{mod } v$. On observe que $S \times_M T$ est une variété et que v et u induisent des éléments v' et u' de $F(S\times_M T)$ localement égaux d'après 1). On a donc $n(v') = f \circ \text{mod } v'$. Et le fait que $S\times_M T \longrightarrow S$ est un épimorphisme permet de conclure.

Rappelons que si M est un module grossier, un élément de F(M) est un objet de Poincaré, si le morphisme associé est l'identité de M.

THEOREME 4.10. *Le morphisme de foncteurs mod (Définition 4.7) fait de M_{c_2} (Définition 3.1) un module grossier pour 'tHS, qui admet un objet de Poincaré si et seulement si c_2 est pair. De plus M est rationnel.*

DEMONSTRATION. Pour appliquer le lemme 4.9 vérifions en les hypothèses. 1) Si $\mathcal{E}'$ et $\mathcal{E}''$ sont deux fibrés 'tHS sur $S \times \mathbb{P}^3$ ayant même morphisme modulaire, on a deux suites exactes

$$0 \longrightarrow \mathcal{E}'^{\vee} \longrightarrow (\pi^*\pi_*\mathcal{E}')^{\vee} \xrightarrow{\alpha'} \mathcal{E}xt^1(\tilde{\mathcal{F}}_{\mathcal{E}'},\mathcal{O}) \longrightarrow 0$$

$$0 \longrightarrow \mathcal{E}''^{\vee} \longrightarrow (\pi^*\pi_*\mathcal{E}'')^{\vee} \xrightarrow{\alpha''} \mathcal{E}xt^1(\tilde{\mathcal{F}}_{\mathcal{E}''},\mathcal{O}) \longrightarrow 0.$$

Le faisceau $\pi_*\mathcal{H}om\,(\mathcal{E}xt^1(\tilde{\mathcal{F}}_{\mathcal{E}'},\mathcal{O}),\ \mathcal{E}xt^1(\tilde{\mathcal{F}}_{\mathcal{E}''},\mathcal{O}))$ est localement libre de rang un d'après le théorème de changement de base. On peut donc, quitte à restreindre S, supposer que $\mathcal{E}xt^1(\tilde{\mathcal{F}}_{\mathcal{E}'},\mathcal{O})$ et $\mathcal{E}xt^1(\tilde{\mathcal{F}}_{\mathcal{E}''},\mathcal{O})$ sont isomorphes. On a $[\alpha'] = [\alpha'']$ et la remarque 3.2.3 permet de conclure que $\mathcal{E}'^{\vee}$ et $\mathcal{E}''^{\vee}$ (et par suite $\mathcal{E}'$ et $\mathcal{E}''$) sont isomorphes.

2) Il suffit d'observer que si $C \longrightarrow B$ est un fibré en coniques, alors $C\times_B C \longrightarrow C$ est un fibré en coniques admettant des fibrés de rang un en tout degré relatif.

Ainsi M est un module grossier. L'affirmation concernant l'objet de Poincaré découle du fait que l'identité de M vérifie la condition de la proposition 4.8 si et seulement si c_2 est pair (On utilise la Proposition 1.1 (3), la Remarque 1.2 et le Théorème 3,4, IV).

La rationalité de M est démontrée dans le Théorème 3.4.,III.

COROLLAIRE 4.11. Le module grossier M(2,3) est rationnel.

DÉMONSTRATION. D'après [6] (Th. 9.7 et sa preuve) tout fibré stable avec $c_1 = 2$, $c_2 = 3$ est un fibré 'tHS.

COROLLAIRE 4.12. Il existe un objet de Poincaré pour $M(2,c_2)$ si et seulement si c_2 est pair.

DÉMONSTRATION. Si c_2 est pair il existe un objet de Poincaré d'après un résultat général de Maruyama [9] (Théorème 6.11). Inversement s'il existe un objet de Poincaré pour $M(2,c_2)$ alors il en existe un pour le module 'tHS correspondant et c_2 est pair d'après le théorème 4.10.

REMARQUE 4.13. En utilisant la Remarque 1.2 et le fait qu'un fibré de 't Hooft correspond à un point lisse dans $M(2,c_2)$ (cf. [6]), on démontre qu'il n'y a pas de fibré de Poincaré sur $U \times \mathbb{P}^3$ où U est un ouvert non-vide de $M(2,c_2)$ dans la composante irréductible contenant les fibrés 'tHS, si c_2 est impair.

5. LE CAS DE $\mathbb{P}^2$.

Dans ce paragraphe nous indiquons brièvement comment notre méthode permet de traiter le cas de $\mathbb{P}^2$.

THÉORÈME 5.1. (*Maruyama-Le Potier*). *Pour qu'il existe un fibré de Poincaré pour* $M(2,c_2)$, *il faut et il suffit que* c_2 *soit pair.*

DEMONSTRATION. La condition suffisante est un cas particulier du résultat général de Maruyama utilisé plus haut.

Pour la condition nécessaire on suppose d'abord que $c_2 \geq 5$. On dit qu'un fibré E (de rang 2) sur $\mathbb{P}^2$ est un fibré de Hulsbergen spécial (HS) s'il existe une conique non-singulière Q_E et un fibré $\mathcal{F}_E$ de rang un sur Q_E tel que E soit l'extension du prolongement $\tilde{\mathcal{F}}_E$ de $\mathcal{F}_E$ à $\mathbb{P}^2$ par un fibré trivial de rang 2. On montre comme dans la proposition 4.3 que Q_E et $\tilde{\mathcal{F}}_E$ sont uniquement déterminés, que $H^o(\mathbb{P}^2,E)$ est de dimension 2, que $\tilde{\mathcal{F}}_E$ est le conoyau de $H^o(\mathbb{P}^2,E) \otimes \mathcal{O} \longrightarrow E$ et que le degré de $\mathcal{F}_E$ est $4-c_2$. On montre ensuite comme dans la proposition 4.4 que cette construction passe bien aux familles de fibrés HS. On observe aussi que $\mathcal{E}xt^1(\tilde{\mathcal{F}}_E, \mathcal{O})$ est le prolongement par zéro d'un fibré de rang un et degré c_2 sur la conique. On définit alors, comme en 4.7, le morphisme mod du foncteur des fibrés HS dans le foncteur représenté par M', variété des $g^1_{c_2}$ sans point de base sur la conique universelle de $\mathbb{P}^2$. On montre comme en 4.8 qu'un morphisme $\varphi : S \longrightarrow M'$ est dans l'image de mod si et seulement s'il existe sur la conique universelle remontée à S un fibré de rang un et degré relatif c_2. On en déduit (comme dans le théorème 4.10) que M' est un module grossier qui admet un fibré de Poincaré si et seulement si c_2 est pair, en utilisant l'exemple 1.5 et la Proposition 2.6.

Comme dans la démonstration de Le Potier, le cas $c_2 = 3$ réclame un traitement spécial. Dans ce cas on sait que le diviseur des droites de saut est une conique de $\mathbb{P}^{2*}$ et que, p et q désignant les projections du diagramme standard

$$\begin{array}{ccc} \mathbb{P}^{2*} & \xleftarrow{\ q\ } & F \\ & & \big\downarrow p \\ & & \mathbb{P}^2 \end{array}$$

$R^1q_* \, p^* E(-1)$ est le prolongement par zéro d'un fibré de rang un et degré un sur cette conique (cf [1], § 6.2). Enfin le module grossier est la variété des coniques non-dégénérées de $\mathbb{P}^{2*}$. On conclut qu'il n'y a pas de fibré de Poincaré parce qu'il n'y a pas de fibré de rang un et degré relatif impair sur la conique universelle.

Remerciements : Ce travail a été effectué à l'Université de Nice. Le second auteur veut remercier l'Université de Nice pour un séjour agréable et d'excellentes conditions de travail.

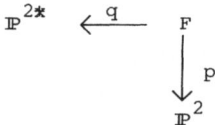

BIBLIOGRAPHIE

[1] W. BARTH — Moduli of vector bundles on the projective plane, Invent. Math. 42 (1977), 63-91.

[2] N. BOURBAKI — Eléments de Mathématiques, Formes sesquilinéaires et formes quadratiques, Hermann, Paris, 1959.

[3] P. DELIGNE — Quadriques, Exposé XII, Groupe de monodromie en géométrie algébrique, S.G.A. VII, Springer Lecture Notes 340 (1977).

[4] G. ELLINGSRUD and S.A. STROMME — On the moduli space of vector bundles of rank two on $\mathbb{P}_3$ with c_1 odd. Preprint Oslo 1979.

[5] R. HARTSHORNE — Algebraic Geometry, Graduate texts in mathematics 52, Springer 1977.

[6] R. HARTSHORNE — Stable vector bundles of rank 2 on $\mathbb{P}^3$, Math. Ann. 238 (1978), 229-280.

[7] K. HULEK — Stable Rank-2 Vector Bundles on $\mathbb{P}_2$ with c_1 Odd, Math. Ann. 242, 241-266 (1979).

[8] J. LE POTIER — Fibrés stables de rang 2 sur $\mathbb{P}_2(\mathbb{C})$, Math. Ann. 241 (1979), 217-256.

[9] M. MARUYAMA — Moduli of stable sheaves II, J. Math. Kyoto Univ. 18 (1978), 557-614.

[10] M.S. NARASIMHAN and S. RAMANAN — Vector bundles on curves, In algebraic Geometry (papers presented in the Bombay Colloquim 1968) (O.U.P. India 1969), 335-346.

[11] P.E. NEWSTEAD Rationality of moduli spaces of stable bundles, Math. Ann. 215 (1975), 251-268. Correction : Math. Ann. 249 (1980), 281-282.

[12] P.E. NEWSTEAD Comparison theorem for conic bundles, Math. Proc. Camb. Phil. Soc. 90 (1981), 21-31.

[13] P.E. NEWSTEAD On the cohomology and the Picard group of a moduli space of bundles on $\mathbb{P}^3$, soumis au Quarterly J. Math.

[14] P.E. NEWSTEAD Pencils on conic bundles. Preprint Liverpool 1982.

[15] S. RAMANAN The moduli spaces of vector bundles over an algebraic curve, Math. Ann. 200 (1973), 69-84.

[16] L. ROTH Algebraic threefolds, Springer, 1955.

[17] J.P. SERRE Espaces fibrés algébriques, Exposé 1, Séminaire Chevalley, 1958.

A. HIRSCHOWITZ
Département de Mathématiques
I.M.S.P.
Parc Valrose - 06034 NICE CEDEX

M.S. NARASIMHAN
Tata Institute of Fundamental
Research,
BOMBAY 400 005 INDE.

FORMULES MULTISECANTES POUR LES COURBES GAUCHES QUELCONQUES

Patrick LE BARZ

Pour une sous-variété X de $\mathbb{P}^N$, un problème classique est de déterminer le cycle des droites k-sécantes à X, dans la grassmannienne G(1,N) des droites de $\mathbb{P}^N$, afin d'obtenir des formules énumératives.

Dans le cas où X est une courbe, il y a essentiellement quatre formules : trois formules trisécantes (k=3) et une formule quadrisécante (k=4). En fonction du degré n et du genre g, elles donnent :

- pour une courbe X de $\mathbb{P}^3$, le nombre t(X) de trisécantes rencontrant une droite fixe,
- pour une courbe X de $\mathbb{P}^3$, le nombre k(X) de tangentes recoupant X,
- pour une courbe X de $\mathbb{P}^4$, le nombre $\Theta(X)$ de trisécantes,
- pour une courbe X de $\mathbb{P}^3$, le nombre q(X) de quadrisécantes.

Ces quatre formules classiques sont

$$\begin{cases} t(X) = \dfrac{(n-1)(n-2)(n-3)}{3} - g(n-2) & \text{(Cayley 1863)} \\ k(X) = 2((n-2)(n-3)+g(n-6)) & \text{(Salmon 1868)} \\ \Theta(X) = \binom{n-2}{3} - g(n-4) & \text{(Berzolari 1895)} \\ q(X) = \frac{1}{12}(n-2)(n-3)^2(n-4) - \frac{1}{2}g(n^2-7n+13-g) & \text{(Cayley 1863)}. \end{cases}$$

Il semble que pendant longtemps les formules aient été appliquées sans trop expliquer leur champ de validité. Par exemple Cayley [2] parlant d'une courbe de bidegré (p,q) située sur une quadrique déclare :

" If p or q exceeds 3, we have the case of a curve through every point where of there can be drawn a line or lines through four or more points and the formula is inapplicable ".

De nombreux auteurs ont donné différentes démonstrations de ces formules ; on peut trouver une bibliographie dans [3], p. 1281-1282. En fait, et cela jusqu'aux démonstrations les plus récentes ([5], [16], [10]), on se restreignait implicitement ou explicitement, à une classe de courbes plus ou moins "génériques".

Dans cet article, on se propose de montrer les formules pour une courbe lisse quelconque, comme cela a été annoncé dans [11]. Cela demande d'abord une définition précise des nombres t, k, Θ ou q associés à X. Cherchons pour fixer les idées, à définir en effet le nombre q(X), pour X courbe dans $\mathbb{P}^3$. On constate malheureusement, par exemple pour une courbe de bidegré (4,4) située sur une quadrique, qu'elle a une infinité de quadrisécantes comme cela a été rappelé plus haut. (On le voit encore plus simplement pour une quartique plane). La solution pour définir q(X) dans tous les cas (ainsi que les autres nombres) est déjà implicite dans Severi [15]. Elle consiste à prendre le nombre d'intersection, dans l'anneau de Chow du schéma de Hilbert formé des quadruplets de $\mathbb{P}^3$, entre d'une part le schéma de Hilbert des quadruplets de X et d'autre part, la sous-variété des quadruplets situés sur une droite.

Avec cette définition, lorsque les deux sous-variétés en question ne se coupent pas proprement (i.e. lorsqu'il y a une infinité de quadrisécantes), le nombre q(X) peut fort bien être <u>négatif</u>. Il vaut par exemple -4 pour la courbe de bidegré (4,4) vue plus haut. Cela signifie dans ce cas que la courbe ne peut être "déformée" en une courbe qui n'aurait qu'un nombre fini de quadrisécantes. (Une courbe avec n=8 et g=9 comme la courbe en question, est en effet nécessairement intersection complète d'une quadrique et d'une quartique).

Maintenant, dans le cas intéressant en pratique où le nombre de quadrisécantes est fini, la formule par définition <u>compte avec multiplicités</u>. Un exemple est donné par la courbe intersection complète d'une surface cubique non-singulière et d'une quintique (n=15, g=31). On trouve alors q=135 ; or toute quadrisécante à la courbe, pour raison de degré, est située sur la surface cubique : il n'y en a donc a priori que 27. Mais chacune d'elles coupe cinq fois la courbe, donc compte cinq fois comme quadrisécante, ce qui explique le 135... Au paragraphe VI on explicite sur les exemples les plus courants comment il faut compter les multiplicités.

Disons maintenant un mot de la démonstration. Au § I, on donne la définition précise de t, k, q et Θ . Au § II, par projection de la courbe C sur un plan générique P, on obtient une courbe C' avec h croisements normaux. La réunion de C' et de h composantes immergées aux points doubles (ce sont les voisinages infinitésimaux des points doubles à l'ordre 1) donne un schéma Γ_o non plan. La courbe C et Γ_o sont en fait

les fibres en 1 et O d'une famille Γ plate sur $\mathbb{C}$. Le schéma de Hilbert relatif $\mathrm{Hilb}^k\,\Gamma/C$ n'est cependant pas plat ; mais si on considère l'ouvert non singulier $\mathrm{Hilb}^k_c\,\mathbb{P}^3$ des k-uplets curvilignes de $\mathrm{Hilb}^k\,\mathbb{P}^3$ (c'est-à-dire situés sur une courbe non-singulière), la trace $\mathrm{Hilb}^k_c\,\Gamma/\mathbb{C}$ est alors plate, d'où l'équivalence rationnelle des cycles :
$[\mathrm{Hilb}^k_c\ C] \sim [\mathrm{Hilb}^k_c\ \Gamma_o]$.

Au § III, on montre que $\mathrm{Hilb}^k_c\,\Gamma_o$ se décompose en $\mathrm{Hilb}^k_c\,C'$ plus d'autres composantes, dues aux composantes immergées de Γ_o. L'évaluation de la contribution de ces composantes dans les formules se fait en IV et V. On obtient les formules sous forme de polynômes en n et h, dont on détermine les coefficients en regardant un certain nombre de cas particuliers, comme par exemple des réunions de droites disjointes.

Enfin au § VI, on montre comment interpréter les formules dans les problèmes de multiplicités rencontrés le plus couramment.

I) $\mathrm{Hilb}^k_c\,\mathbb{P}^N$, $A\ell^k\,\mathbb{P}^N$ et formules k-sécantes

II) Les schémas Γ et Γ_o

III) Les composantes de $(\mathrm{Hilb}^k_c\,\Gamma_o)_{red}$ et la classe de $\mathrm{Hilb}^k C$

IV) Trisécantes

V) Quadrisécantes

VI) Comment compter les multiplicités.

I. $\mathrm{Hilb}^k_c\,\mathbb{P}^N$, $A\ell^k\,\mathbb{P}^N$ ET FORMULES k-SECANTES

a. Le corps de base est $\mathbb{C}$. On appelle k-uplet un schéma ξ de dimension O et longueur k, c'est-à-dire $\dim_{\mathbb{C}}\Gamma(\xi,\mathcal{O}_\xi) = k$. Si X est un schéma projectif, on dénote par $\mathrm{Hilb}^k X$ le schéma de Hilbert [6] des k-uplets de X. Un élément de $(\mathrm{Hilb}^k X)_{red}$ est donc un idéal I de $\mathcal{O}_X$ avec Supp $\mathcal{O}_X/I$ fini et $\dim_{\mathbb{C}}\Gamma(X,\mathcal{O}_X/I) = k$.

Rappelons qu'un sous-schéma Y de XxT, fini sur T, est plat sur T si et seulement si la longueur des fibres Y_t est une constante k indépendante de $t \in T$. Le schéma $\mathrm{Hilb}^k X$ est solution d'un problème universel en ce sens que la donnée de la situation précédente est équivalente à la donnée d'un morphisme

$$f : T \longrightarrow \mathrm{Hilb}^k X.$$

Dans ce qui suit, on se placera dans $\mathrm{Hilb}^k \mathbb{P}^N$. Remarquons qu'un élément de $\mathrm{Hilb}^k \mathbb{P}^1$ est donné, à une constante près, par un polynôme de degré k ; donc $\mathrm{Hilb}^k \mathbb{P}^1$ est Proj $\Gamma(\mathbb{P}^1, \mathcal{O}_{\mathbb{P}^1}(k))$, lui-même isomorphe à $\mathbb{P}^k$.

Plus généralement, pour une courbe C non-singulière, on sait [6] que $\mathrm{Hilb}^k C$ est isomorphe à $\mathrm{Sym}^k C$, le produit symétrique de C. En fait, pour une variété quelconque X, on a toujours un morphisme

$$\pi : \mathrm{Hilb}^k X \longrightarrow \mathrm{Sym}^k X$$

(résultant du foncteur "Hilb $\longrightarrow$ Chow"), mais ce n'est pas toujours un isomorphisme, ni même une modification. Par contre, si l'on se restreint à l'ouvert (non dense en général) $\mathrm{Hilb}^k_{\neq} X$ formé des k-uplets de points distincts, on a un isomorphisme sur l'image $\mathrm{Sym}^k_{\neq} X$ dans $\mathrm{Sym}^k X$.

Maintenant, si X/S est un schéma relatif sur S, on a la notion de schéma de Hilbert relatif $\mathrm{Hilb}^k X/S$ sur S, dont la fibre en s est $\mathrm{Hilb}^k(X_s)$.

b. Les notions qui suivent ont été introduites dans [12] et [13] où on développait les détails techniques nécessaires.

Un k-uplet ξ de $\mathbb{P}^N$ est dit <u>curviligne</u> si au voisinage de chacun des points de son support, il est situé sur une courbe non-singulière. L'ouvert $\mathrm{Hilb}^k_c \mathbb{P}^N$ des k-uplets curvilignes est non-singulier de dimension Nk, mais non dense en général dans $\mathrm{Hilb}^k \mathbb{P}^N$. (L'ouvert $\mathrm{Hilb}^k_{\neq} \mathbb{P}^N$ est par contre dense dans $\mathrm{Hilb}^k_c \mathbb{P}^N$).

Un k-uplet ξ de $\mathbb{P}^N$ est dit <u>aligné</u> s'il est sous-schéma d'une droite (réduite), appelée <u>axe</u> du k-uplet. L'ensemble des k-uplets alignés est une sous-variété notée $\mathrm{Al}^k \mathbb{P}^N$, de $\mathrm{Hilb}^k_c \mathbb{P}^N$. Le morphisme "Axe" de $\mathrm{Al}^k \mathbb{P}^N$ dans la grassmannienne $G(1,N)$, qui à un k-uplet aligné associe l'unique droite sur laquelle il est situé, est une fibration de fibre-type $\mathrm{Hilb}^k \mathbb{P}^1 \simeq \mathbb{P}^k$. On a donc $\dim \mathrm{Al}^k \mathbb{P}^N = 2N-2+k$.
Cette fibration provient d'un fibré vectoriel de rang k+1 et, par le théorème de Hirsch-Leray, on peut donner explicitement des générateurs de son anneau de Chow $A^{\bullet}(\mathrm{Al}^k \mathbb{P}^N)$. (Voir proposition 8).

c. Soit X une sous-variété de $\mathbb{P}^N$ et $\mathrm{Hilb}^k_c X$ la trace de $\mathrm{Hilb}^k X$ sur l'ouvert $\mathrm{Hilb}^k_c \mathbb{P}^N$.

Considérons le diagramme

$$\begin{array}{ccc} Al^k\, \mathbb{P}^N & \xrightarrow{i} & \mathrm{Hilb}_c^k\, \mathbb{P}^N \\ & & \uparrow \\ & & \mathrm{Hilb}_c^k\, X \end{array}$$

où i est l'injection canonique.

Par définition, <u>le cycle des droites k-sécantes à X</u> est $i^*[\mathrm{Hilb}_c^k\, X]$ dans l'anneau de Chow $A^\bullet(Al^k\, \mathbb{P}^N)$.

(On note par [ ] le cycle associé à un sous-schéma).

Commençons par le cas le plus simple : définir <u>le nombre de trisécantes d'une courbe de</u> $\mathbb{P}^4$.

DEFINITION 1. *Soit C une courbe de* $\mathbb{P}^4$. *Considérons le diagramme, où les dimensions sont indiquées entre parenthèses :*

$$\begin{array}{lcccl} (9) & Al^3\, \mathbb{P}^4 & \xrightarrow{i} & \mathrm{Hilb}_c^3\, \mathbb{P}^4 & (12) \\ & & & \uparrow & \\ & & & \mathrm{Hilb}_c^3\, C & (3)\,. \end{array}$$

Le "nombre de trisécantes à C" est par définition le degré $\Theta(C)$ *du 0-cycle* $i^*\,[\mathrm{Hilb}_c^3\, C]$ *dans* $A^\bullet(Al^3\, \mathbb{P}^4)$.

Pour les courbes de $\mathbb{P}^3$ par contre, le cycle des trisécantes $i^*\,[\mathrm{Hilb}_c^3\, C]$ est un 1-cycle :

$$\begin{array}{lcccl} (7) & Al^3\, \mathbb{P}^3 & \xrightarrow{i} & \mathrm{Hilb}_c^3\, \mathbb{P}^3 & (9) \\ & & & \uparrow & \\ & & & \mathrm{Hilb}_c^3\, C & (3)\,. \end{array}$$

Pour parler du degré d'un 0-cycle, on doit donc évaluer $i^*\,[\mathrm{Hilb}_c^3\, C]$ sur un élément de $A^1(Al^3\, \mathbb{P}^3)$.

Soit G(1,3) la grassmannienne des droites de $\mathbb{P}^3$ et soit $\sigma \in A^1(G(1,3))$ le cycle de Schubert des droites coupant une droite fixe. Comme on a un morphisme

$$\mathrm{Axe} : Al^3\, \mathbb{P}^3 \longrightarrow G(1,3)$$

le cycle $\mathrm{Axe}^*\sigma$ est dans $A^1(Al^3\, \mathbb{P}^3)$. On donne donc la

DEFINITION 2. *Soit C une courbe de* $\mathbb{P}^3$. *Le "nombre de trisécantes à C recoupant une droite fixe" est par définition le degré t(C) du 0-cycle*

$$i^* [\mathrm{Hilb}^3_c\ C]\ .\ \mathrm{Axe}^*\sigma$$

dans $A^\cdot(\mathrm{Al}^3\ \mathbb{P}^3)$.

Maintenant, considérons dans $\mathrm{Hilb}^3\ \mathbb{P}^1 \simeq \mathbb{P}^3$ le complémentaire de $\mathrm{Hilb}^3_{\neq}\ \mathbb{P}^1$: c'est la surface D_o des triplets non simples, i.e. non formés de trois points distincts. Cette surface singulière est la fibre-type d'une sous-fibration D de $\mathrm{Al}^3\ \mathbb{P}^3$: D est formée des triplets alignés non simples. Comme le cycle associé [D] est dans $A^1(\mathrm{Al}^3\ \mathbb{P}^3)$, on donne la

DEFINITION 3. *Soit C une courbe de* $\mathbb{P}^3$. *Le "nombre de tangentes à C recoupant C "est par définition le degré k(C) du 0-cycle*

$$i^* [\mathrm{Hilb}^3_c\ C]\ .\ [D]$$

dans $A^\cdot(\mathrm{Al}^3\ \mathbb{P}^3)$.

Enfin, pour définir le nombre q(C) de quadrisécantes à une courbe de $\mathbb{P}^3$, on regarde le diagramme

$$\begin{array}{llcll} (8) & \mathrm{Al}^4\ \mathbb{P}^3 & \xrightarrow{\ i\ } & \mathrm{Hilb}^4_c\ \mathbb{P}^3 & (12) \\ & & & \uparrow & \\ & & & \mathrm{Hilb}^4_c\ C & (4) \end{array}$$

et on constate que les dimensions sont complémentaires. On donne donc la

DEFINITION 4. *Soit C une courbe de* $\mathbb{P}^3$. *Le "nombre de quadrisécantes à C" est par définition le degré q(C) du 0-cycle* $i^* [\mathrm{Hilb}^4_c\ C]$.

Dans les paragraphes qui suivent, on va démontrer que ces quatre nombres t, k, Θ et q ont bien la forme annoncée dans l'introduction, pour C courbe lisse quelconque.

II. LES SCHEMAS Γ ET Γ_o.

1. Projection générique et Définition de Γ .

Dans tout ce qui suit, C est une courbe lisse de $\mathbb{P}^3$, de degré n avec h "points doubles apparents". Plus précisément, on choisit un plan P de $\mathbb{P}^3$ et un point ω tel que la projection de C sur P par ω soit "générique", c'est-à-dire :

i) La courbe image C' possède exactement h points doubles ordinaires $(M_j)_{j=1,2..h}$, aucun des points M_j n'étant d'inflexion pour les deux branches de C' en M_j et d'autre part, les deux tangentes en M_j sont chacune transverses en n-2 autres points.

ii) Au dessus de $C' - \{M_j\}_{j=1,2..h}$, la projection de C par ω est un isomorphisme, alors que l'image réciproque de chaque M_j est formée de deux points.

REMARQUE 1. Si C est une courbe irréductible, il est bien connu que son genre g est lié au nombre h par :

$$g = \frac{(n-1)(n-2)}{2} - h.$$

D'après ([9], prop. 1.4) ou encore ([8], III, exemple 9.8.3), il existe un sous-schéma réduit et irréductible Γ de $\mathbb{P}^3 \times \mathbb{C}$, plat sur $\mathbb{C}$ avec

$$\Gamma_1 = C \quad \text{et} \quad (\Gamma_o)_{red} = C'.$$

Prenons pour cela un système de coordonnées homogènes $(x : y : z : t)$ pour lequel P a pour équation $\{z=0\}$, le point ω étant $(0 : 0 : 1 : 0)$. Le schéma Γ a alors la structure réduite de l'adhérence, dans $\mathbb{P}^3 \times \mathbb{C}^*$, de l'image de $C \times \mathbb{C}^*$ par le plongement

$$j : C\times\mathbb{C}^* \longrightarrow \mathbb{P}^3\times\mathbb{C}^*$$

$$(x : y : z : t),\lambda \longmapsto (x : y : \lambda z : t),\lambda .$$

Le sous-schéma Γ_o a des nilpotents qui "pointent" hors de P aux h points doubles de C'. (Voir dessin dans [8] , p. 260).
Dans ce qui suit, on va préciser la structure de Γ_o.

2. Rappel sur le Schéma $\mathscr{X}$.

On a défini dans ([12], II) le schéma relatif $\mathscr{X}/\mathbb{C}$ de la façon suivante. Dans $\mathbb{C}^3_{x,y,z} \times \mathbb{C}_\lambda$, on considère $\mathscr{X}$ défini par la réunion des deux plans d'équations

$$\{x = z+\lambda = 0\} \text{ et } \{y = z -\lambda = 0\} .$$

L'idéal de $\mathscr{X}$ est alors $J = (x,z+\lambda) \cap (y,z-\lambda)$.
C'est encore
$J = (x,z+\lambda).(y,z-\lambda) = (xy,\ x(z-\lambda),\ y(z+\lambda),\ z^2-\lambda^2)$

car x, y, z+λ et z-λ sont des coordonnées de $\mathbb{C}^4$. La fibre $\mathcal{X}_\lambda$ dans $\mathbb{C}^3$ est alors la réunion des deux droites disjointes $\begin{cases} x=0 \\ z=-\lambda \end{cases}$ et $\begin{cases} y=0 \\ z=\lambda \end{cases}$, tandis que la fibre $\mathcal{X}_o$ a pour idéal dans $\mathcal{O}_{\mathbb{C}^3}$:

$$J_o = (xy,\ xz,\ yz,\ z^2).$$

Si Q désigne le quadruplet d'idéal m^2 où m est l'idéal de l'origine dans $\mathbb{C}^3$, on a l'égalité de sous-schémas de $\mathbb{C}^3$:

$$\mathcal{X}_o = (\mathcal{X}_o)_{red} \cup Q$$

puisque $J_o = (xy,z) \cap m^2$; le schéma $(\mathcal{X}_o)_{red}$ est la réunion des deux axes de coordonnées dans le plan {z=0}.

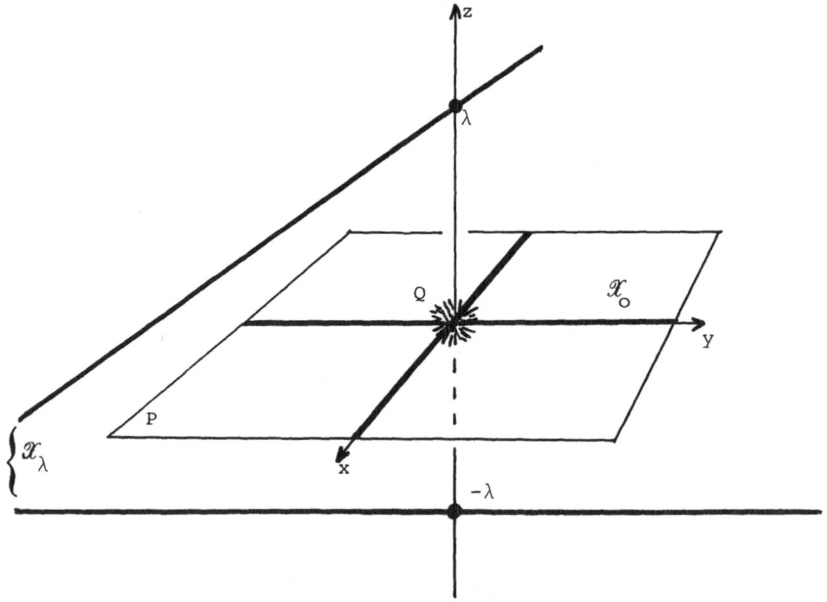

On a montré dans [12] les résultats essentiels suivants :

i. Tout k-uplet curviligne dans $\mathcal{X}_o$ est limite de k-uplets curvilignes dans $\mathcal{X}_o$ avec longueur au plus 2 en chaque point du support (loc. cit. prop. 2).

ii. Tout k-uplet curviligne dans $\mathcal{X}_o$ est limite de k-uplets curvilignes dans $\mathcal{X}_\lambda$ avec λ≠0 (loc. cit. prop. 3).

iii. $\mathrm{Hilb}^2\ \mathcal{X}_o$ est génériquement réduit (loc. cit. prop. 4) .

3. Description de Γ_o.

PROPOSITION 1. *Pour tout point double M de C', le germe (Γ,M) est isomorphe, comme schéma relatif sur le germe $(\mathbb{C},0)$ au germe $(\mathcal{X},0)$.*

En particulier, on a un isomorphisme des germes (Γ_o,M) et $(\mathcal{X}_o,0)$.

PREUVE : Soit ω le point de $\mathbb{P}^3$ par lequel s'effectue la projection sur P, a et b les deux points de C ayant même image M dans P. Les tangentes T_aC et T_bC sont disjointes puisque ω a été choisi en 1) pour que C' n'ait que des points doubles ordinaires comme singularités. Considérons le plan H déterminé par les trois points ω , $P\cap T_aC$ et $P\cap T_bC$.
Soit (x, y, z) un système de coordonnées de $\mathbb{P}^3$-H avec

$$M = (0,0,0) \qquad a = (0,0,-1) \qquad b = (0,0,1)$$

$$P = \{z = 0\} \qquad \overline{\omega M} = \{x=y=0\},$$

et les projections de T_aC et T_bC sur P étant respectivement les droites $\{x=0\}$ et $\{y=0\}$.
La projection sur P par ω est alors exprimée par $(x,y,z) \longmapsto (x,y)$.

On a des représentations paramétriques de C (dans les voisinages formels de a et b) données par

$$\begin{cases} x = \varphi(t) \\ y = t \\ z = -1+O(t^3) \end{cases} \qquad \text{et} \qquad \begin{cases} x = t \\ y = \psi(t) \\ z = 1+O(t^3) \end{cases}$$

où φ,ψ sont dans $\mathbb{C}[[t]]$ et val φ , val $\psi \geq 2$.
D'autre part, on a dit plus haut que Γ est l'adhérence dans $\mathbb{P}^3\times\mathbb{C}$ de l'image du plongement

$$j : C\times\mathbb{C}^* \longrightarrow \mathbb{P}^3\times\mathbb{C}^*$$
$$(x:y:z:t),\lambda \longmapsto (x:y:\lambda z:t),\lambda .$$

Soit J la partie de Γ formée des $(0:0:\lambda:1),\lambda$ et $(0:0:-\lambda:1),\lambda$. C'est la réunion de deux sections de $\mathbb{P}^3\times\mathbb{C}$ au dessus de $\mathbb{C}$. Soit Ω le voisinage formel de J dans $\mathbb{P}^3\times\mathbb{C}$ et Ω^* sa trace sur $\mathbb{P}^3\times\mathbb{C}^*$. Le schéma $\Gamma\cap\Omega^*$ est défini par la réunion de deux sous-schémas disjoints A et B, images de Spec $\mathbb{C}[[t]] \times \mathbb{C}^*$ par

$$(t,\lambda) \longmapsto \begin{cases} x = \varphi(t) \\ y = t \\ z = -\lambda(1+O(t^3)) \end{cases} \quad \text{et} \quad (t,\lambda) \longmapsto \begin{cases} x = t \\ y = \psi(t) \\ z = \lambda(1+O(t^3)). \end{cases}$$

L'idéal dans $\mathcal{O}_\Omega$ de l'adhérence $\bar{A}$ et $\bar{B}$ de chacun de ces sous-schémas est respectivement

$$\mathcal{A} = (\varphi(y)-x,\ z+\lambda(1+O(y^3)))$$

et $$\mathcal{B} = (\psi(x)-y,\ z-\lambda(1+O(x^3))).$$

Or Γ est défini comme ayant la structure réduite sur l'adhérence de A U B, soit $\bar{A} \cup \bar{B}$. L'idéal d'une réunion étant l'intersection des idéaux, l'idéal de Γ dans $\mathcal{O}_\Omega$ est $\mathcal{A}\cap\mathcal{B}$.

Dans $\mathbb{C}[[x,y,z,\lambda]]$, on fait le changement de variables

$$\begin{cases} X = x - \varphi(y) \\ Y = y - \psi(x) \end{cases} \qquad \begin{cases} z = z \\ \lambda = \lambda \end{cases}$$

et on pose $\tilde{u}(X,Y) = 1+O(y^3)$, $\tilde{v}(X,Y) = 1+O(x^3)$.

L'idéal $\mathcal{A}\cap\mathcal{B}$ du germe $(\Gamma,0)$ dans $\mathbb{C}^4$ devient après ce changement

$$(X,z+\lambda u(Y)) \cap (Y,z-\lambda v(x))$$

où $u(Y) = \tilde{u}(0,Y)$ et $v(X) = \tilde{v}(X,0)$.

(On a donc $u(0) = v(0) = 1$). Or on a le lemme évident suivant :

<u>LEMME 1</u>. Dans $\mathbb{C}^4$, soit H_1 et H_2 deux hyperplans se coupant suivant un plan π . Soit φ_1 (resp. φ_2) un automorphisme de H_1(resp. H_2) laissant π fixe point par point. Alors φ_1 et φ_2 se recollent en un automorphisme φ de $H_1 \cup H_2$.

On applique ce lemme à $\mathbb{C}^4 = \mathbb{C}^4_{X,Y,z,\lambda}$. Soit H_1(resp. H_2) l'hyperplan $\{X=0\}$ (resp. $\{Y=0\}$).

Soit $\varphi_1 : (0,Y,z,\lambda) \longmapsto (0,Y,z-\lambda+\lambda u(Y),\lambda)$

$\varphi_2 : (X,0,z,\lambda) \longmapsto (X,0,z+\lambda-\lambda v(X),\lambda)$.

Comme $u(0) = v(0) = 1$, on voit que

$$\varphi_1(0,0,z,\lambda) = (0,0,z,\lambda) = \varphi_2(0,0,z,\lambda).$$

D'où un automorphisme φ de $H_1 \cup H_2$ envoyant Γ sur le sous-schéma défini par l'idéal

$$(X,z+\lambda) \cap (Y,z-\lambda),$$

et qui bien sûr commute avec la projection sur $\mathbb{C}_\lambda$. Cet idéal n'est rien d'autre que $(XY,X(z-\lambda),Y(z+\lambda),z^2-\lambda^2)$, c'est-à-dire l'idéal de $\mathcal{X}$. La proposition 1 est donc démontrée.

REMARQUE 2. Si m_i est l'idéal de M_i dans $\mathcal{O}_{\mathbb{P}^3}$ et Q_i le quadruplet d'idéal m_i^2, on a l'égalité de sous-schémas

$$\Gamma_o = C' \cup Q_1 \cup Q_2 \cup \ldots \cup Q_h$$

puisque on a l'égalité des idéaux

$$I(\Gamma_o) = I(C') \cap m_1^2 \cap m_2^2 \cap \ldots \cap m_h^2$$

dans $\mathcal{O}_{\mathbb{P}^3}$.

III. LES COMPOSANTES DE $\mathrm{Hilb}^k_c \Gamma_o$ ET LA CLASSE DE $\mathrm{Hilb}^k_c C$.

1. Equivalence Rationnelle dans $\mathrm{Hilb}^k_c \mathbb{P}^N$.

Donnons tout d'abord la

DEFINITION 5. *Un sous-schéma X d'un schéma est réduit en tant que cycle si pour toute composante irréductible Y de dimension maximum de X_{red}, X est génériquement réduit le long de Y.*

La proposition suivante est la base de la démonstration des formules multisécantes.

PROPOSITION 2. *Soit H un schéma ; soit $\bar{U}$ un ouvert de $\mathbb{P}^1$ contenant 0 et 1 et soit $U = \bar{U} - \{0,1\}$.*

Soit Z un sous-schéma réduit de $H \times U$, plat sur U. On désigne l'adhérence de Z dans $H \times \bar{U}$ par $\bar{Z}$.

Si les fibres $\bar{Z}_0$ et $\bar{Z}_1$ sont réduites en tant que cycles dans H, alors on a l'équivalence rationnelle dans H :

$$[(\bar{Z}_0)_{red}] = [(\bar{Z}_0)] \sim [(\bar{Z}_1)] = [(\bar{Z}_1)_{red}] .$$

PREUVE. D'après ([9], prop.1.4) et ([7], I.9.5.9 et I.9.5.10), l'adhérence $\bar{Z}$ est l'unique sous-schéma relatif plat sur $\bar{U}$ dont la restriction à U et Z. (De plus $\bar{Z}$ est réduit). Donc $\bar{Z}_o$ et $\bar{Z}_1$ sont les fibres d'un morphisme plat à valeurs dans $\mathbb{P}^1$; par définition même de l'équivalence rationnelle, on a $[\bar{Z}_o] \sim [\bar{Z}_1]$ dans H.

Mais vu l'hypothèse faite sur $\bar{Z}_o$ et $\bar{Z}_1$ on a

$[\bar{Z}_j] = [(\bar{Z}_j)_{red}]$, $j=0,1$.

D'où le résultat.

REMARQUE 3. Dans la suite, on appliquera la proposition 2 où $H = \mathrm{Hilb}_c^k \mathbb{P}^N$. Soit alors $X/\bar{U}$ un sous-schéma relatif de $\mathbb{P}^N x\bar{U}$ et soit $\mathrm{Hilb}^k X/U$ le sous-schéma relatif correspondant de $\mathrm{Hilb}^k \mathbb{P}^N xU$. On note $\mathcal{Z}$ la trace $\mathrm{Hilb}_c^k X/U = \mathrm{Hilb}^k X/U \cap (\mathrm{Hilb}_c^k \mathbb{P}^N xU)$.

Par exemple si $X/\bar{U}$ est une courbe relative, lisse sur U, on a

$$\mathrm{Sym}^k X/U = \mathrm{Hilb}^k X/U = \mathrm{Hilb}_c^k X/U = \mathcal{Z}$$

qui est réduit et lisse (donc plat) sur U. On pourra donc appliquer la proposition 2.

On remarque par ailleurs que le schéma relatif $\mathrm{Hilb}_c^k X/\bar{U}$ est un sous-schéma (fermé) de $\mathrm{Hilb}_c^k \mathbb{P}^N x\bar{U}$, par définition même, qui contient $\mathcal{Z} = \mathrm{Hilb}_c^k X/U$. On a donc, par définition de l'adhérence, l'inclusion $\bar{\mathcal{Z}} \subset \mathrm{Hilb}_c^k X/\bar{U}$ et c'est une inclusion de schémas, puisque $\mathcal{Z}$ est réduit.

On en déduit l'inclusion (de schémas) entre les fibres :

$$\bar{\mathcal{Z}}_j \subset \mathrm{Hilb}_c^k X_j \qquad j=0,1.$$

2. Triplets.

Dans tout ce qui suit, on fait N=3 et dans ce paragraphe, k=3.

Soit $C \subset \mathbb{P}^3$ une courbe lisse et Γ_o le sous-schéma de $\mathbb{P}^3$ introduit en II. On rappelle que $(\Gamma_o)_{red}$ est une projection C' de C sur un plan générique P de $\mathbb{P}^3$. On va détailler les différentes composantes de $(\mathrm{Hilb}_c^3 \Gamma_o)_{red}$.

DEFINITION 6. *Pour $j=1,2\ldots h$, notons $\bar{W}_j$ l'adhérence dans $(\mathrm{Hilb}_c^3 \Gamma_o)_{red}$ de la partie W_j formée des triplets $t = m \cup d$ où*

- d est un doublet de Γ_o de support $\{M_j\}$

- m est un point de C' - $\{M_j\}$

Remarquons que $\bar{W}_j$ (de dimension 3) n'est pas irréductible si C' ne l'est pas.

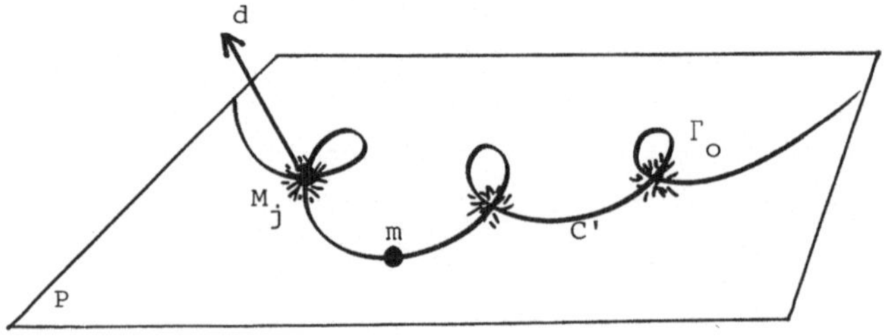

PROPOSITION 3. *$(Hilb_c^3\ \Gamma_o)_{red}$ est réunion de $(Hilb_c^3 C')_{red}$ et des $\bar{W}_j$ $(j=1,2..h)$.*

PREUVE. Soit t un triplet curviligne contenu dans Γ_o. Si t est contenu dans P, comme $P \cap \Gamma_o = C'$, on a $t \in (Hilb_c^3 C')_{red}$. Si maintenant $t \not\subset P$, le support de t ne peut être formé de trois points distincts. Il contient donc forcément un des points M_j, car en dehors de ces points on a $\mathcal{O}_{\Gamma_o}| = \mathcal{O}_{C'}|$. D'autre part, toujours puisque $t \not\subset P$, la longueur de t en l'un des M_j est strictement plus grande que 1.

Si le support de t est réduit $\{M_j\}$, d'après la proposition 1 et (II.2.i), on a $t \in \bar{W}_j$. Sinon, Supp t = $\{M_j, m\}$ où $m \neq M_j$. Les longueurs de t aux points M_j et m sont 2 et 1 ; ainsi $t \in W_j$.

3. Quadruplets.

Ici, k=4 ; comme précédemment, on va détailler les composantes de $(Hilb_c^4\ \Gamma_o)_{red}$.

DEFINITION 7. *Pour $j=1,2...h$, notons $\bar{U}_j$ l'adhérence dans $(Hilb_c^4\ \Gamma_o)_{red}$ de la partie U_j formée de quadruplets $q = d \cup d'$ où*

- *d est un doublet de Γ_o de support $\{M_j\}$*
- *d' est un doublet contenu dans $C' - \{M_j\}$ (et non $\Gamma_o - \{M_j\}$!).*

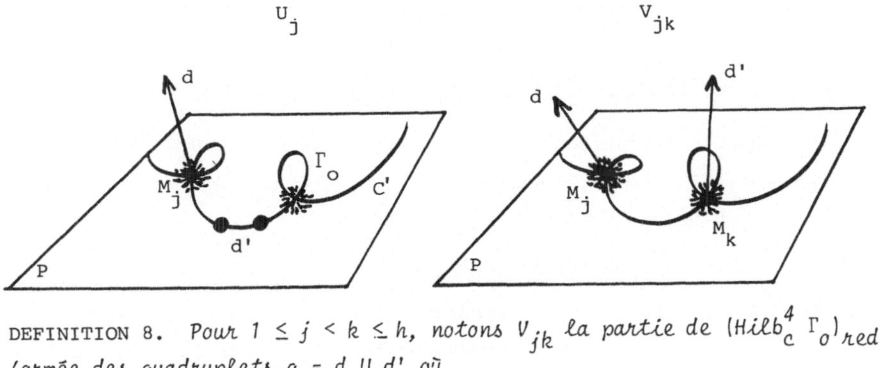

DEFINITION 8. *Pour $1 \leq j < k \leq h$, notons V_{jk} la partie de $(Hilb_c^4\ \Gamma_o)_{red}$ formée des quadruplets $q = d \cup d'$ où*

- *d est un doublet de Γ_o de support $\{M_j\}$*
- *d' est un doublet de Γ_o de support $\{M_k\}$.*

Remarquons que $\bar{U}_j$ et V_{jk} sont de dimension 4.

PROPOSITION 4. *$(Hilb_c^4\ \Gamma_o)_{red}$ est réunion de $(Hilb_c^4\ C')_{red}$, des $\bar{U}_j$ et des V_{jk}.*

PREUVE. Soit q un quadruplet curviligne contenu dans Γ_o. Si q est contenu dans P, on a $q \in (Hilb_c^4 C')_{red}$ puisque $P \cap \Gamma_o = C'$. Si $q \not\subset P$, le support de q ne peut-être formé de quatre points distincts. Il contient forcément un des points M_i car, en dehors de ces points, on a $\mathcal{O}_{\Gamma_o}| = \mathcal{O}_{C'}|$. D'autre part, toujours puisque $q \not\subset P$, la longueur de q en l'un des M_j est strictement plus grande que 1.

Premier cas : Supp q = $\{M_j, m, m'\}$ où m et m' appartiennent à $C' - \{M_j\}$ et $m \neq m'$. Les longueurs de q en ces trois points sont 2, 1 et 1 et donc $q \in U_j$, avec $d' = \{m,m'\}$ dans la définition 7.

Deuxième cas : Supp q = $\{M_j, m\}$ où $m \in C' - \{M_j\}$.
Si les longueurs de q en ces deux points sont 2 et 2,

- soit $q \in U_j$ si $m \neq M_\ell$ pour tout ℓ ,
- soit $q \in V_{jk}$ si $m = M_k$.

Si les longueurs de q en ces mêmes points sont 3 et 1, à cause de la proposition 1 et (II.2.i), on a $q \in \bar{U}_j$.

Troisième cas : Supp q = $\{M_j\}$. Pour la même raison que précédemment $q \in \bar{U}_j$.

4. Application à $Hilb_c^3 C$ et $Hilb_c^4 C$.

a. On pose (notations de la proposition 2) : $\bar{U} = \mathbb{C}$, $U = \bar{U} - \{0,1\}$. Pour le sous-schéma relatif Γ de $\mathbb{P}^3 \times \bar{U}$ défini en (II.1), on pose

$$Z = Z^k = Hilb_c^k \Gamma/U \text{ pour } k=3,4.$$

LEMME 2 : On a $(\bar{Z}_o)_{red} = (Hilb_c^k \Gamma_o)_{red}$.

PREUVE : D'après la fin de la remarque 2, on a déjà une inclusion $\bar{Z}_o \subset Hilb_c^k\ \Gamma_o$, d'où l'inclusion correspondante entre les réduits associés.

Cherchons alors à montrer l'inclusion $(Hilb_c^k\ \Gamma_o)_{red} \subset (\bar{Z}_o)_{red}$. Puisque $(Hilb_c^k\ \Gamma_o)_{red}$ est d'une part réduit et d'autre part contenu dans la fibre en 0, il suffit donc de montrer l'inclusion

$$(Hilb_c^k\ \Gamma_o)_{red} \subset \bar{Z}.$$

Soit alors ξ_o un k-uplet curviligne contenu dans Γ_o. On se ramène évidemment au cas où Supp ξ est l'un des points doubles M_j.
Comme d'après la proposition 1, le germe (Γ, M_j) est isomorphe à $(\mathcal{X}, 0)$,

on a d'après (II.2.ii) $\xi_o = \lim \xi_\lambda$ avec ξ_λ dans $(\mathrm{Hilb}_c^k \Gamma_\lambda)_{red}$ et $\lambda \neq 0$. Donc $\xi_o \in \bar{Z}$ par définition même de Z.

b. Un corollaire immédiat de ce lemme et des propositions 3 et 4 est la

PROPOSITION 5. *On a l'égalité des cycles*

$$[(\bar{Z}_o^3)_{red}] = [(\mathrm{Hilb}_c^3 C')_{red}] + \sum_{1 \le j \le h} [\bar{W}_j] \text{ dans } \mathrm{Hilb}_c^3 \mathbb{P}^3,$$

$$[(\bar{Z}_o^4)_{red}] = [\mathrm{Hilb}_c^4 C')_{red}] + \sum_{1 \le j \le h} [\bar{U}_j] + \sum_{1 \le j < k \le h} [V_{jk}]$$

dans $\mathrm{Hilb}_c^4 \mathbb{P}^3$.

Pour pouvoir appliquer cette proposition et conclure, il ne reste qu'à montrer le

LEMME 3 : Le sous-schéma $\bar{Z}_o$ est réduit en tant que cycle dans $\mathrm{Hilb}_c^k \mathbb{P}^3$.

PREUVE : D'après ([1] , corollaire 7), pour une courbe plane C', le schéma de Hilbert $\mathrm{Hilb}^k C'$ est réduit. En fait, on n'a besoin que de beaucoup moins, à savoir que $\mathrm{Hilb}_c^k C'$ est génériquement réduit, ce qui peut se voir de manière élémentaire.

Par ailleurs, $\mathrm{Hilb}_c^3 \Gamma_o$ est génériquement réduit le long de W_j, car avec les notations de la définition 6, $\mathrm{Hilb}_c^2 \Gamma_o$ est génériquement réduit au voisinage de d (vu la proposition 1 et II.2 iii).
Pour les mêmes raisons, $\mathrm{Hilb}_c^4 \Gamma_o$ est génériquement réduit le long de U_j et V_{jk}.

c. On sait que la fibre Γ_1 de Γ en 1 est la courbe C, donc $\bar{Z}_1 = \mathrm{Hilb}^k C$ est réduit en tant que cycle, puisque c'est une sous-variété non-singulière de $\mathrm{Hilb}_c^k \mathbb{P}^3$. Les propositions 2 et 5 et le lemme 3 permettent donc d'énoncer la

PROPOSITION 6. *On a les équivalences rationnelles*

$$[\mathrm{Hilb}_c^3 C] \sim [(\mathrm{Hilb}_c^3 C')_{red}] + \sum_{1 \le j \le h} [\bar{W}_j] \text{ dans } \mathrm{Hilb}_c^3 \mathbb{P}^3$$

$$[\mathrm{Hilb}_c^4] \sim [(\mathrm{Hilb}_c^4 C')_{red}] + \sum_{1 \le j \le h} [\bar{U}_j] + \sum_{1 \le j < k \le h} [V_{jk}]$$

dans $\mathrm{Hilb}_c^4 \mathbb{P}^3$.

IV. TRISECANTES

Soit C une courbe de $\mathbb{P}^3$. Considérons comme toujours le diagramme

$$\begin{array}{llll} (7) & Al^3\,\mathbb{P}^3 & \xrightarrow{i} & Hilb_c^3\,\mathbb{P}^3 \quad (9) \\ & & & \uparrow \\ & & & Hilb_c^3\,C \quad (3) \end{array}$$

où les dimensions sont entre parenthèses.

La proposition qui suit est essentielle dans tout ce paragraphe.

PROPOSITION 7 : *Pour C courbe de $\mathbb{P}^3$ de degré n avec h points doubles apparents, pour tout cycle z dans A^1 $(Al^3\,\mathbb{P}^3)$, le degré du 0-cycle*

$$z.\, i^*[\,Hilb_c^3\,C\,]$$

est de la forme

$$n(n-2)(an+b) + h(cn+d)$$

où a, b, c, d dans $\mathbb{Q}$ ne dépendent que de z.

Cette proposition concernant les courbes de $\mathbb{P}^3$ ne sera démontrée qu'en 2). Nous allons voir auparavant qu'elle entraîne comme corollaire la formule des trisécantes pour une courbe de $\mathbb{P}^4$.

1. Trisécantes d'une Courbe de $\mathbb{P}^4$.

a. Soit $\mathbb{P}^3$ un hyperplan de $\mathbb{P}^4$ et considérons le diagramme commutatif où les flèches sont les injections canoniques :

$$\begin{array}{lcccl} (9) & Al^3\,\mathbb{P}^4 & \xrightarrow{j} & Hilb_c^3\,\mathbb{P}^4 & (12) \\ & \uparrow & & \uparrow u & \\ (7) & Al^3\,\mathbb{P}^3 & \xrightarrow{i} & Hilb_c^3\,\mathbb{P}^3 & (9)\ . \end{array}$$

Ensemblistement, dans $Hilb_c^3\,\mathbb{P}^4$, on a

$$Al^3\,\mathbb{P}^4 \cap Hilb_c^3\,\mathbb{P}^3 = Al^3\,\mathbb{P}^3.$$

Autrement dit, $Al^3\,\mathbb{P}^4$ et $Hilb_c^3\,\mathbb{P}^4$ ne se coupent pas proprement dans $Hilb_c^3\,\mathbb{P}^4$, comme on le voit en examinant les dimensions. D'après le théorème de Fulton-Mac Pherson [4], leur intersection en tant que cycle peut être choisie à support dans $Al^3\,\mathbb{P}^3$. C'est-à-dire qu'il existe un cycle $z \in A^1(Al^3\,\mathbb{P}^3)$ tel que

$$i_* z = u^*\, j_*\,[Al^3\,\mathbb{P}^4].$$

b. Soit alors C une courbe de $\mathbb{P}^4$. La projection sur un hyperplan générique est un isomorphisme ; fixons un tel hyperplan, noté $\mathbb{P}^3$, et soit $\hat{C}$ la projection de C sur $\mathbb{P}^3$.

D'après [8], ex. 9.8.3, on a dans $\mathbb{P}^4 \times \mathbb{C}$ un sous-schéma relatif trivial avec fibre C en 1 et $\hat{C}$ en 0. On a donc dans $\mathrm{Hilb}^3_c \mathbb{P}^4 \times \mathbb{C}$ un sous-schéma relatif trivial avec fibre $\mathrm{Hilb}^3_c C$ en 1 et $\mathrm{Hilb}^3_c \hat{C}$ en 0. Ainsi, $[\mathrm{Hilb}^3_c C]$ est rationnellement équivalent à $[\mathrm{Hilb}^3_c \hat{C}]$ dans $\mathrm{Hilb}^3_c \mathbb{P}^4$. On en conclut :

$$\Theta(C) = \deg j^* [\mathrm{Hilb}^3_c C] = \deg j^* [\mathrm{Hilb}^3_c \hat{C}] = \Theta(\hat{C}).$$

D'après a) et la formule des projections, on a pour W sous-variété de $\mathrm{Hilb}^3_c \mathbb{P}^3$:

$$j_*(j^* u_* [W]) = u_* [W] \cdot j_* [\mathrm{Al}^3 \mathbb{P}^4] =$$

$$u_*([W] \cdot u^* j_* [\mathrm{Al}^3 \mathbb{P}^4]) = u_*([W] \cdot i_* \mathbb{Z})$$

$$= u_* i_* (i^* [W] \cdot \mathbb{Z}).$$

En particulier pour $W = \mathrm{Hilb}^3_c \hat{C}$, on a

$\Theta(\hat{C}) = \deg j^* [\mathrm{Hilb}^3_c \hat{C}] = \deg j_* j^* (u_* [\mathrm{Hilb}^3_c \hat{C}]) = \deg(i^* [\mathrm{Hilb}^3_c \hat{C}] . \mathbb{Z})$.

D'après la proposition 7, on a donc

$$\Theta(C) = \Theta(\hat{C}) = n(n-2)(an+b) + h(cn+d),$$

puisque $\hat{C}$ a évidemment même degré et même nombre de points doubles apparent que C.

c. Les courbes particulières suivantes n'ont pas de trisécante dans $\mathbb{P}^4$:

- une cubique rationnelle située dans un $\mathbb{P}^3$;
  d'où $(n=3,\ h=1)$: $\Theta = 3(3a+b) + 3c + d = 0$;
- deux coniques disjointes situées dans des plans transverses ; d'où $(n=4,\ h=4)$: $\Theta = 4(8a+2b+4c+d) = 0$.

Par ailleurs l'expression générale de $\Theta(C)$ montre que pour une droite, $\Theta = -(a+b)$. Or trois droites dans $\mathbb{P}^4$ ont une trisécante, d'où dans ce cas $\Theta = -3(a+b)+1$. (Un calcul rapide de multiplicité montre en effet que cette trisécante ne compte qu'une seule fois).

Plus généralement, pour n droites de $\mathbb{P}^4$, on a

$$\Theta = -n(a+b) + \binom{n}{3}.$$

Ceci donne, puisqu'alors $h = \binom{n}{2}$, l'identité :

$$-\,n(a+b) + \binom{n}{3} = n(n-2)(an+b) + \binom{n}{2}(cn+d).$$

On en tire les deux équations :

$$\begin{cases} -a-b = \dfrac{2c+d}{2} \\[2mm] \dfrac{1}{6} = a + \dfrac{c}{2} \end{cases} .$$

Jointes aux deux équations obtenues précédemment, on a un système inversible, d'où l'on tire les valeurs cherchées :

$$a = -\frac{1}{3}, \quad b = \frac{4}{3}, \quad c = 1, \quad d = -4 .$$

On a ainsi démontré le

<u>THEOREME 1</u> : Soit C une courbe lisse de $\mathbb{P}^4$, de degré n avec h points doubles apparents. Alors le nombre de trisécantes est

$$\Theta(C) = (n-4)h - \frac{n(n-2)(n-4)}{3} .$$

En particulier si C est une courbe irréductible de genre g, on a

$$\Theta(C) = \binom{n-2}{3} - g(n-4).$$

2. <u>Formule Trisécante Générale dans $\mathbb{P}^3$</u>.

Il s'agit maintenant de prouver la proposition 7. D'après la proposition 6, pour tout cycle $\mathcal{Z}$ dans $A^1(Al^3\,\mathbb{P}^3)$, on a

$$\deg(\mathcal{Z}.i^*[\mathrm{Hilb}^3_c C]) = \deg(\mathcal{Z}.i^*[\mathrm{Hilb}^3_c C']) + \sum_{1 \leq j \leq h} \deg(\mathcal{Z}.i^*[\overline{W}_j]).$$

a. Evaluation de $\deg(\mathcal{Z}.i^*[\mathrm{Hilb}^3_c C'])$.

Soit P le plan contenant C' et regardons le diagramme commutatif où les flèches sont les injections canoniques :

$$\begin{array}{lccccl} (7) & Al^3\,\mathbb{P}^3 & \xrightarrow{\ i\ } & \mathrm{Hilb}^3_c\,\mathbb{P}^3 & & (9) \\ & \uparrow & & u\uparrow & & \\ (5) & Al^3\,P & \xrightarrow{\ i_1\ } & \mathrm{Hilb}^3_c\,P & & (6)\ . \end{array}$$

On a $Al^3P = Al^3\,\mathbb{P}^3 \cap \mathrm{Hilb}^3_c P$ et l'intersection n'est pas propre vu les dimensions. D'après le théorème de Fulton-Mac Pherson cité plus haut, pour tout $\mathcal{Z}$ dans $A^{\bullet}(Al^3\,\mathbb{P}^3)$, il existe $K \in A^{\bullet}(Al^3P)$ tel que $i_{1*}K = u^*i_*\mathcal{Z}$.

Soit W une sous-variété de $Hilb^3_c P$. En utilisant plusieurs fois la formule des projections comme en 1°), on arrive à

$$i_*(i^* u_* [W].\mathcal{Z}) = u_* [W].i_* \mathcal{Z} = u_*([W].u^* i_* \mathcal{Z})$$

$$= u_*([W].i_{1*} K) = u_*(i_1^* [W].K).$$

En particulier pour $W = Hilb^3_c C'$, on a

$$\deg(i^* [Hilb^3_c C'] .\mathcal{Z}) = \deg(i_1^* [Hilb^3_c C'] .K).$$

Or l'évaluation des formules trisécantes pour une courbe plane a été faite dans [13]. D'après la proposition 3 de loc. cit. on a

(*) $\quad \deg i_1^* [Hilb^3_c C'] . K = n(n-2)(an+b)$

où a et b sont dans $\mathbb{Q}$.

Cette proposition se prouvait en évaluant $i_1^* [Hilb^3_c C']$ sur une base de $A^2(Al^3 P)$.

C'est le même genre de technique que nous allons utiliser en b), pour

b. Evaluation de $\deg(\mathcal{Z}.i^* [W_j])$.

Comme $Axe : Al^k \mathbb{P}^N \longrightarrow G = G(1,N)$ est une fibration de fibre $\mathbb{P}^k$ provenant d'un fibré vectoriel, le théorème de Hirsch-Leray a pour conséquence (indirecte) la proposition suivante, démontrée dans [13], qui donne une base de $A^{\bullet}_{\mathbb{Q}}(Al^k \mathbb{P}^N)$.

(On note $A_{\mathbb{Q}}$ pour $A \otimes_{\mathbb{Z}} \mathbb{Q}$).

PROPOSITION 8. *Soit $\mathcal{H}_i$ dans $\mathbb{P}^N$ une hypersurface fixée de degré i. Pour $k \geq i$, on note $[\bar{H}_i^k]$ la classe dans $A^i(Al^k \mathbb{P}^N)$ du cycle adhérence de la sous-variété H_i^k formée des k-uplets alignés ξ vérifiant*

$$long\ (\xi \cap (Axe\ \xi \cap \mathcal{H}_i)) \geq i,$$

c'est-à-dire, dit plus simplement que les k-uplets de H_i^k ont "<u>au moins i points sur $\mathcal{H}_i$</u>".

Alors on a l'égalité dans $A^{\bullet}_{\mathbb{Q}}(Al^k \mathbb{P}^N)$ des sous-espaces vectoriels

$$A^i_{\mathbb{Q}}(Al^k \mathbb{P}^N) \text{ et } \bigoplus_{0 \leq j \leq i} Axe^* A^{i-j}_{\mathbb{Q}}(G).[\bar{H}_j^k].$$

Appliquons cette proposition à $A^1(Al^3 \mathbb{P}^3)$. Nous voyons qu'une base de $A^1_{\mathbb{Q}}(Al^3 \mathbb{P}^3)$, comme espace vectoriel, est donnée par

- $Axe^* \sigma$ où σ est le cycle de Schubert de G(1,3) formé des droites rencontrant une droite fixe (σ est une base de $A^1(G(1,3))$;
- $[\bar{H}_1^3]$, cycle des triplets alignés "dont au moins un point est sur un plan fixé".

On a le

LEMME 4 .

α) $\deg \mathrm{Axe}^*\sigma \,.\, i^*[\bar{W}_j] = n-2.$

β) $\deg [\bar{H}_1^3] \,.\, i^*[\bar{W}_j] = n$.

PREUVE : α) C'est un exemple de calcul de multiplicités (ici 1), comme on en rencontrera au paragraphe VI.

Considérons une droite δ non contenue dans P ; la sous-variété $\mathrm{Al}^3_\delta\, \mathbb{P}^3$ de $\mathrm{Al}^3\, \mathbb{P}^3$ formée de triplets dont l'axe rencontre δ , représente le cycle $\mathrm{Axe}^*\sigma$. Nous allons montrer que les sous-variétés W_j et $\mathrm{Al}^3_\delta\, \mathbb{P}^3$ se coupent transversalement en n-2 éléments dans $\mathrm{Hilb}^3_C\, \mathbb{P}^3$.

On suppose que l'intersection p de δ avec P n'est situé sur aucune des tangentes aux points doubles de C'. Soit $M = M_j$ correspondant à $\bar{W}=\bar{W}_j$.

L'intersection au sens des schémas, de la droite $\overline{pM}$ et de C' est formée de n-2 points simples et d'un doublet d de support $\{M\}$. L'intersection (ensembliste) de $\bar{W}$ et $\mathrm{Al}^3_\delta\, \mathbb{P}^3$ est donc formée de n-2 triplets réunion de d et de l'un des n-2 autres points simples.

Soit t un de ces triplets et choisissons un système de coordonnées inhomogènes (x,y,z) d'un ouvert de carte de $\mathbb{P}^3$ pour lequel P est donné par $\{z=0\}$, M par (0,0,0) et p par (1,1,0). On a vu (proposition 1) que dans $\mathbb{C}[[x,y,z]]$, l'idéal de Γ_o est donné par

$$((x+O(y^2))(y+O(x^2)),\ z(x+O(y^2)),\ z(y+O(x^2)),z^2).$$

La droite $\overline{pM}$ coupe Γ_o au voisinage de M en le doublet d d'idéal $(x^2,x-y,z)$. Une carte de $\mathrm{Hilb}^2\, \mathbb{P}^3$ en d correspond à l'idéal

$$(x^2+ax+b,\ -y+x+cx+d,\ -z+ex+f).$$

D'autre part, t est donné par la réunion de d et du point simple N, par exemple (-1, -1, 0). Le germe de C' en N est donné par $y=-1+\varphi(x+1)$ avec $\varphi(t)\in C[[t]]$ et $\varphi'(0)\neq 1$ puisque $\overline{pM}$ coupe C' transversalement en dehors de M.

Ainsi une carte de $\mathrm{Hilb}^3\, \mathbb{P}^3$ en t est donnée par a, b, c, d, e, f, u, v, w où (u, v, w) paramètrent les points (-1+u, -1+v, w) voisins de N.

<u>Expression de W dans cette carte</u> : les doublets de support $\{M\}$ qui sont intersection de Γ_o et de la droite d'équations

$$y = x + cx \qquad\qquad z = ex$$

ont pour idéal $(x^2, -y+x+cx, -z+ex)$ car 1+c est inversible. D'où les équations de $\bar{W}$:

$$a = b = d = f = 0 \qquad \text{(pour le doublet)}$$

$$w = 0,\quad v = \varphi(u) \qquad \text{(pour le point simple).}$$

<u>Expression de $Al^3_\delta\, \mathbb{P}^3$ dans cette carte</u> : comme la droite δ n'est pas contenue dans P, une représentation paramétrique en est : $\lambda \longmapsto (1+\lambda\alpha,\ 1+\lambda\beta,\lambda)$. On voit qu'elle a un point commun avec la droite Δ d'équations $\{-y+x+cx+d = -z+ex+f = 0\}$ si et seulement si l'on a la relation

$$-1 - \beta\left(\frac{e+f}{1-\alpha e}\right) + (1+c)\left(1 + \alpha\,\frac{e+f}{1-\alpha e}\right) + d = 0.$$

La relation linéaire tangente à l'origine en est $(\alpha-\beta)(e+f)+c+d=0$. D'autre part, dire que le point $(-1+u, -1+v,w)$ appartient à la droite Δ s'exprime par

$-v+u+c(-1+u)+d = -w+e(-1+u)+f = 0.$

Les neuf relations linéaires tangentes ainsi obtenues forment un système de rang maximal, car $\varphi'(0) \neq 1$.

Ceci prouve donc l'assertion de transversalité et par suite la partie α) du lemme 3.

β) Laissé au lecteur : calcul analogue de transversalité dans des coordonnées locales pour voir que chacun des n triplets compte avec multiplicité 1.

c. Conclusion : Comme $Axe^*\sigma$ et $[\bar{H}^3_1]$ forment une base de $A^1_{\mathbb{Q}}(Al^3\,\mathbb{P}^3)$ on conclut par le lemme 3 que pour $z \in A^1(Al^3\,\mathbb{P}^3)$, l'intersection $z.i^*[\bar{W}_j]$ a un degré de la forme cn+d.

D'après (*), la somme

$$\deg(z.i^*\,[Hilb^3_c C']) + \sum_{1\leq j\leq h} \deg\,(z.i^*\,[\bar{W}_j])$$

est de la forme $n(n-2)\,(an+b) + h(cn+d)$, ce qui prouve la proposition 7.

3 <u>Application à k(C) et t(C).</u>

Soit C une courbe lisse de $\mathbb{P}^3$.

a. Cherchons le nombre de tangentes qui recoupent C. On a posé (définition 3) $k(C) = \deg\, i^*\,[Hilb^3_c\, C]\ .\ [D]$. D'après la proposition 7, k(C) est donc de la forme $n(n-2)(an+b)+h(cn+d)$.

Pour une droite, on trouve $k = -(a+b)$. Pour une réunion de n droites disjointes, il n'y a évidemment pas d'autres tangentes qui recoupent que les droites elles-mêmes, d'où $k = -n(a+b)$.
Comme dans ce cas $h = \binom{n}{2}$, on a l'identité

$$-n(a+b) = n(n-2)(an+b) + \binom{n}{2}(cn+d).$$

On en déduit déjà deux équations :

$$-(a+b) = -2b - \frac{d}{2} \text{ et } 0 = a + \frac{c}{2}.$$

Par ailleurs, une cubique gauche n'a pas de trisécante, d'où a fortiori pas de tangente qui la recoupe. Donc $k=0$, ce qui donne (ici $n=3$ et $h=1$) la troisième équation :

$$0 = 3(3a+b) + 3c + d.$$

Enfin, regardons la courbe C réunion d'une droite et d'une conique disjointes. Il y a évidemment deux tangentes à C recoupant C, plus celles venant de la droite. Donc $k(C) = 2-(a+b)$.
(Il faut vérifier que chacune des deux tangentes compte avec multiplicité 1 ce qui se fait comme dans le lemme 3). Cela donne donc l'équation (ici $n=3$ et $h=2$) :

$$2 - (a+b) = 3(3a+b)+2(3c+d).$$

Les quatre équations qui précèdent forment un système inversible d'où l'on tire les valeurs

$$a = 1, \quad b = -5, \quad c=-2, \quad d=12.$$

On a ainsi montré le

THEOREME 2 : Soit C une courbe lisse de $\mathbb{P}^3$, de degré n, avec h points doubles apparents. Alors le nombre de tangentes qui recoupent est

$$k(C) = n(n-2)(n-5)-2h(n-6).$$

En particulier, si C est une courbe irréductible de genre g on a

$$k(C) = 2((n-2)(n-3) + g(n-6).$$

REMARQUE 4 : Si C est singulière, ce nombre est à interpréter ; voir au paragraphe VI.

b. Cherchons le nombre de trisécantes à C, coupant une droite fixe. On a posé (définition 2) $t(C) = \deg i^* [\mathrm{Hilb}_C^3 C]. \mathrm{Axe}^*\sigma$ où σ est le cycle de Schubert des droites coupant une droite fixe.

D'après la proposition 7 (plus exactement le lemme 3(α)), le nombre t(C) est de la forme

$$n(n-2)(an+b)+h(n-2) .$$

Les deux courbes suivantes n'ont pas de trisécante rencontrant une droite fixe :

- une droite ; comme n=1, h=0, on a

$$0 = -(a+b)$$

- une cubique gauche ; comme n=3, h=1, on a

$$0 = 3(3a+b)+1.$$

On en tire $a = -\frac{1}{6}$, $b = \frac{1}{6}$ et on peut donc énoncer le

<u>THEOREME 3</u> : Soit C une courbe lisse de $\mathbb{P}^3$, de degré n, avec h points doubles apparents. Alors le nombre de trisécantes rencontrant une droite fixe est

$$t(C) = h(n-2) - \binom{n}{3}.$$

En particulier, si C est une courbe irréductible de genre g, on a

$$t(C) = 2\binom{n-1}{3} - g(n-2).$$

V. QUADRISECANTES

Soit C une courbe lisse de $\mathbb{P}^3$, de degré n avec h points doubles apparents. D'après la définition 4, le nombre de quadrisécantes q(C) est $\deg i^* [\mathrm{Hilb}_c^4 C]$. D'après la proposition 6, on a

$$q(C) = \deg i^* [\mathrm{Hilb}_c^4 C'] + \sum_{1 \leq j \leq h} \deg i^* [\overline{U}_j] + \sum_{1 \leq j < k \leq h} \deg i^* [V_j k]$$

avec les notations des définitions 7 et 8.

a. Evaluation de $\deg i^* [\mathrm{Hilb}_c^4 C']$

Elle se fait comme en (IV.2.a) : on se ramène à la proposition 4 de [13] qui montre que le degré est de la forme

$$n(n-2)(n-3)(an+b)$$

avec a et b dans $\mathbb{Q}$.

b. Evaluation de deg $i^* [\bar{U}_j]$.

C'est la partie pénible de la démonstration. Elle a été faite dans [13] à titre d'exercice. Ce degré est de la forme

$\alpha+\beta n+\gamma n^2$.

On peut également utiliser la méthode plus simple décrite dans [11] , où l'on se contentait de montrer que ce degré ne dépend que de n. On peut alors conclure, en d), de la même manière.

c. Evaluation de deg $i^* [V_{jk}]$.

Elle se fait comme dans le lemme 3, par un calcul de transversalité. Nous allons montrer que ce degré est 1 en prouvant que $Al^4 \mathbb{P}^3$ et V_{jk} se coupent transversalement en un seul élément de $Hilb_c^4 \mathbb{P}^3$.

Soient M_j et M_k les points de C' correspondant à la composante V_{jk} (définition 8). D'après (II.1.i), la droite $\overline{M_jM_k}$ n'est pas tangente à C', ni en M_j, ni en M_k. L'intersection ensembliste $Al^4 \mathbb{P}^3 \cap V_{jk}$ est $\{\xi\}$ où ξ est le quadruplet situé sur $\overline{M_j M_k}$, ayant longueurs 2 en M_j et M_k. Considérons un système inhomogène (x,y,z) d'un ouvert de carte de $\mathbb{P}^3$ contenant M_j et M_k. On peut prendre P donné par $\{z=0\}$, M_j par (0,0,0) et M_k par (1,0,0). Le quadruplet ξ est donné par la réunion des deux doublets d'idéal (x^2,y,z) et $((x-1)^2,y,z)$. D'où une carte de $Hilb_c^4 \mathbb{P}^3$ en ξ donnée par

$$\begin{cases} (x^2+ax+b,\ y+a'x+b',\ z+a''x+b'') \\ ((x-1)^2+cx+d,\ y+c'x+d',\ z+c''x+d''). \end{cases}$$

L'expression de $Al^4 \mathbb{P}^3$ dans cette carte est alors immédiate, les deux doublets devant avoir même axe :

$a'=c'$, $b'=d'$ et $a''=c''$, $b''=d''$.

Les doublets de support $\{M_j\}$ s'expriment dans la carte précédente par

$a=b=0$, $b'=0$, $b''=0$

et ceux de support $\{M_k\}$ par

$c=d=0$, $c'+d'=0$, $c''+d''=0$.

On obtient ainsi les huit équations de V_{jk}.

Les douze équations qui précédent sont indépendantes, ce qui prouve l'assertion de transversalité.

d. La Formule.

De ce qui précède résulte que q(C) est de la forme

$$n(n-2)(n-3)(an+b) + h(\alpha+\beta n+\gamma n^2) + \binom{h}{2}.$$

En particulier, pour une droite, on a q=2(a+b). Or on sait ([14], p. 301) que pour quatre droites disjointes générales de $\mathbb{P}^3$, il y a exactement deux droites qui les rencontrent toutes les quatre. (Considérer la quadrique déterminée par trois d'entre elles). Pour la courbe formée de la réunion de ces quatre droites, on a donc q=8(a+b)+2, puisqu'un quadruplet aligné sur cette courbe est ou bien situé sur l'une des quatre droites, ou bien l'un des deux quadruplets précédents. (On vérifie, toujours par un calcul rapide de transversalité, que la multiplicité est 1 : voir au paragraphe VI).

Plus généralement, pour la courbe formée de n droites disjointes générales de $\mathbb{P}^3$, on a

$$q = 2n(a+b) + 2\binom{n}{4}.$$

Ceci conduit à l'identité, puisqu'ici $h = \binom{n}{2}$:

$$2n(a+b)+2\binom{n}{4} = n(n-2)(n-3)(an+b)+\binom{n}{2}(\alpha+\beta n+\gamma n^2) + \binom{\binom{n}{2}}{2}.$$

On en tire les trois relations :

$$2(a+b) - \frac{1}{2} = 6b - \frac{1}{2}\alpha + 1/4$$

$$2(a+b) = \frac{1}{2}\alpha + \beta + 2\gamma$$

$$\frac{1}{12} = a + \frac{\gamma}{2} + \frac{1}{8}$$

(obtenues pour n=0, n=2, n=∞).

Par ailleurs, une cubique gauche n'a pas de trisécante, donc a fortiori, pas de quadrisécante ; comme ici n=3, h=1, on a

$$0 = \alpha+3\beta+ 9\gamma.$$

Enfin, la réunion de deux coniques possède une seule quadrisécante (comptant une seule fois comme on le voit par un calcul simple).
On a donc, comme n=4, h=4 :

$$1=8(4a+b)+4(\alpha+4\beta+16\gamma) + 6.$$

Les cinq équations qui précèdent sont indépendantes et on trouve les valeurs :

$$a = -\frac{1}{24},\ b = \frac{13}{24},\ \alpha = 6,\ \beta = -2,\ \gamma = 0.$$

On a donc démontré le

<u>THEOREME 4</u> : Soit C une courbe lisse de $\mathbb{P}^3$, de degré n, avec h points doubles apparents. Alors le nombre de quadrisécantes est

$$q(C) = -\frac{n}{24}(n-2)(n-3)(n-13) - 2h(n-3) + \binom{h}{2}.$$

En particulier, si C est irréductible de genre g, on a

$$q(C) = \frac{1}{12}(n-2)(n-3)^2(n-4) - \frac{1}{2}g(n^2-7n+13-g).$$

VI CALCULS DE MULTIPLICITES

1. Cas où les Multiplicités sont toutes 1.

Soit C une courbe lisse de $\mathbb{P}^3$. On a montré dans ([10], p.127) que si la courbe C est "générique", la seule multiplicité à considérer dans t(C), k(C) et q(C) est 1.

Dans [10] "générique" signifiait précisément :

i) pour toute droite coupant C en trois points simples, les trois tangentes ne sont pas dans un même plan.

ii) pour toute droite coupant C en quatre points simples, si les quatre tangentes sont deux à deux disjointes, alors aucune d'elles n'est tangente à la quadrique définie par les quatre autres.

iii) pour toute tangente en x coupant C en y, le plan osculateur $Osc_x C$ est transverse à $T_y C$.

iv) il n'y a ni quintisécante, ni bitangente, ni tangente stationnaire, ni tangente recoupant C deux fois.

Cependant, des cas dégénérés comme une bitangente peuvent compter une fois en tant que quadrisécante (mais alors quatre fois comme tangente qui recoupe...). Voir le paragraphe suivant.

2. Multiplicité d'une Tangente stationnaire dans k(C) ; autres exemples.

a. Supposons qu'au voisinage d'un point, la courbe C soit donnée en coordonnées affines (x,y,z) par

$$y = x^3+x^4\,\alpha(x) \qquad\qquad z = x^4+x^5\,\beta(x)$$

où α et β sont définies dans un voisinage de 0 dans $\mathbb{C}$. C'est le cas le plus simple de tangente stationnaire. <u>Nous allons voir qu'une telle tangente compte avec multiplicité 2 dans k(C)</u>.

Le triplet t d'idéal $I_0 = (x^3,y,z)$ est contenu dans C et une carte de $\mathrm{Hilb}^3_C\,\mathbb{P}^3$ en t est donnée par

$$I = (x^3+ax^2+bx+c,\ y+a'x^2+b'x+c',\ z+a''x^2+b''x+c'').$$

Dans cette carte, $\mathrm{Al}^3\,\mathbb{P}^3$ est donné par

(1)(2) $\qquad a'=a''=0.$

De plus, l'hypersurface D de $\mathrm{Al}^3\,\mathbb{P}^3$ est donnée par l'annulation du discriminant de x^3+ax^2+bx+c, soit :

$$(3)\qquad \begin{vmatrix} 0 & 0 & b & c & 0 \\ 0 & b & 2a & b & c \\ b & 2a & 3 & a & b \\ 2a & 3 & 0 & 1 & a \\ 3 & 0 & 0 & 0 & 1 \end{vmatrix} = 0$$

Maintenant, comment s'exprime Hilb^3C au voisinage de t ? On doit dire qu'un triplet est situé sur C si et seulement si on a l'inclusion renversée d'idéaux $I\supset J$ où $J = (-y+x^3+x^4\alpha(x),\ -z+x^4+x^5\beta(x))$ est l'idéal de C. C'est équivalent à dire $I+J = I$. Or

$$I+J = I + (x^4\alpha(x)+x^3+a'x^2+b'x+c',\ x^5\beta(x)+x^4+a''x^2+b''x+c'').$$

On doit donc avoir les deux égalités dans $\mathcal{O}(V)$ où V est voisinage de 0 dans $\mathbb{C}$:

$$\begin{cases} x^4\alpha(x)+x^3+a'x^2+b'x+c' = f(x)(x^3+ax^2+bx+c) \\ x^5\beta(x)+x^4+a''x^2+b''x+c'' = g(x)(x^3+ax^2+bx+c) \end{cases}$$

où $f,g\in\mathcal{O}(V)$ dépendent de a,b,c,a',b',c',a",b",c".

Désignons l'idéal maximal de $\mathbb{C}\{a,b,c,a',b',c',a'',b'',c''\}$ par m. On écrira par abus $\varphi=\psi+m^i$ pour $\varphi-\psi\in m^i$.

Par identification des développements, si

$$f(x) = \Sigma f_i x^i \quad\text{et}\quad g(x) = \Sigma g_i x^i,$$

on a, pour la première égalité :

$$\begin{cases} c' = f_o c \\ b' = f_o b + f_1 c \\ a' = f_o a + f_1 b + f_2 c \\ 1 = f_o + f_1 a + f_2 b + f_3 c \\ \alpha_o = f_1 + f_2 a + f_3 b + f_4 c \\ \alpha_1 = f_2 + f_3 a + f_4 b + f_5 c \end{cases}$$

Les trois dernières équations montrent

$f_o = 1+m, \quad f_1 = \alpha_o+m, \quad f_2 = \alpha_1+m$.

En reportant dans les trois premières, on a

(4) $c'=c+m^2$ (5) $b'=b+\alpha_o c+m^2$

et, vu $a'=0$

(6) $0 = a+\alpha_o b+\alpha_1 c+m^2$.

L'identification analogue pour la deuxième égalité donne

(7) $c''=m^2$ (8) $b''=c+m^2$

(9) $b + \beta_o c = m^2$.

L'idéal engendré dans $\mathbb{C}\{a,b,c,a',b',c',a'',b'',c''\}$ par ces neuf relations est alors

$(a+\lambda c+m^2, a', a'', b+\beta_o c+m^2, b'+\mu c+m^2, b''-c+m^2, c'-c+m^2, c''+m^2, 9c^2+m^3)$

où λ, β_o, μ sont des constantes.

Cet idéal est donc de longueur 2 : c'est la multiplicité avec laquelle on doit compter dans $k(C)$ une tangente stationnaire "simple".

b. Le cas qui vient immédiatement après est celui d'une tangente stationnaire correspondant à la paramétration locale

$y = x^3+\ldots \qquad z = x^5+\ldots$

c'est un calcul analogue laissé au lecteur devoir que la tangente qui recoupe compte ici dans $k(C)$ avec multiplicité 3.

De même, à une paramétration

$y = x^4+\ldots \qquad z = x^5+\ldots$

correspond une tangente "superstationnaire" qui compte avec multiplicité 6 dans $k(C)$ comme tangente qui recoupe.

c. Donnons rapidement sans calcul, quelques autres exemples de multiplicités.

- Il résulte de la définition même de q(C) que si une droite d coupe C en p points et si les tangentes en quatre points d'intersection quelconques sont en position générale, alors d compte $\binom{p}{4}$ fois comme quadrisécante, puisque cela correspond à $\binom{p}{4}$ quadruplets alignés comptant chacun une fois.

- Une tangente d à C en m_o qui recoupe C en m_1 compte une fois dans k(C) si $T_{m_1}C$ est transverse au plan $Osc_{m_o}C$.

Si $Tm_1C \subset Osc_{m_o}C$, d compte deux fois dans k(C), à condition que $Osc\ m_1C \neq Osc\ m_oC$.

En effet, reprenons les notations de ([10], I.2.b) et soit (puisque w=0)

$$y=vx_1 + \ldots \qquad z=\alpha x_1^2 + \ldots \qquad (\alpha\neq 0)$$

les équations de C au voisinage de m_1. On note ici D ce qui est noté Al_{21} dans loc. cit. L'idéal de $D \cap Hilb^3_c C$ au voisinage du triplet correspondant à d est

$$(\alpha x_1^2, vx_1 + \frac{\psi''(0)}{2}\ a,\ b,\ a',\ b',\ a'',\ b'',\ y_1,\ z_1)$$

et il est de longueur 2 puisque $\psi''(0) \neq 0$ (m_o non stationnaire) et $\alpha \neq 0$ (transversalité de $Osc_{m_1}C$).

- Regardons le cas d'une tangente stationnaire recoupant simplement la courbe. Si on désigne par $\bullet\!\!\longrightarrow$ un doublet et par $\bullet\!\!\overset{3}{\longrightarrow}$ un triplet aligné, l'intersection est donc $\bullet\!\!\overset{3}{\longrightarrow}\ \bullet$

On peut vérifier que cette droite va (en général) compter 1 fois en tant que quadrisécante.

Cependant en tant que tangente qui recoupe, elle comptera 6 fois dont

2 fois pour l'intersection $\bullet\!\!\overset{3}{\longrightarrow}$ (paragraphe a)

et 4 fois pour l'intersection $\bullet\!\!\longrightarrow\ \bullet$

- Enfin, pour le cas d'une bitangente, où l'intersection est

$$\bullet\!\!\longrightarrow \quad \bullet\!\!\longrightarrow$$

on a des résultats analogues : cette droite compte (en général) une fois en tant que quadrisécante, mais 4 fois en tant que tangente qui recoupe, dont

2 fois pour l'intersection $\bullet\!\!\longrightarrow\ \bullet$

et 2 fois pour l'intersection $\bullet\ \ \bullet\!\!\longrightarrow$.

3. Extension au cas singulier ; cusps ordinaires.

Si l'on observe bien les démonstrations données dans ce travail, on voit qu'elles s'étendent à des courbes réduites assez générales de $\mathbb{P}^3$.

Soit C une courbe vérifiant

(Z) $\left\{\begin{array}{l}\text{C est une courbe réduite et pour tout point x de C, on a} \\ \dim T_x C \leq 2 \text{ (tangent de Zariski).}\end{array}\right.$

Alors sa projection sur un plan générique est isomorphe sur l'image C', en dehors d'un certain nombre h de points doubles. Les conditions (II.1.i et ii) sont donc vérifiées. On définit alors comme précédemment un schéma relatif plat $\Gamma/\mathbb{C}$ avec fibres Γ_1=C et Γ_o égal à C' plus des composantes immergées aux h points doubles.

De plus, en (III.4.c), on utilisait le fait que $\mathrm{Hilb}^k C$ est lisse, donc réduit, si C est lisse. Mais d'après le corollaire 7 de [1] , $\mathrm{Hilb}^k C$ est réduit si C vérifie seulement l'hypothèse (Z) précédente.

Pour une courbe de $\mathbb{P}^3$ vérifiant (Z), les trois formules donnant t(C), k(C) et q(C) sont donc valables.

Seulement, pendant tout l'article, on a utilisé les synonymes

trisécante	pour triplet aligné
quadrisécante	pour quadruplet aligné
tangente qui recoupe	pour triplet aligné non simple.

Pour une courbe singulière, les deux premières assimilations sont légitimes, mais pas la troisième.

Regardons en effet le cas le plus simple d'une paramétration

$$x = t^2 + \dots \qquad y = t^3 + \dots \qquad z = t^4 + \dots$$

en coordonnées affines. On voit que toutes les droites passant par l'origine et contenues dans le plan Oxy, coupent C avec multiplicité 2 en O. On ne saurait pour autant dire que ce sont des tangentes à C.

En particulier, pour obtenir une formule pour "le nombre de tangentes à C qui recoupent C", on doit retrancher de k(C) le nombre de droites passant par un cusp p et contenues dans le plan osculateur P à C en p, qui recoupent C.

Et ce, en comptant les multiplicités. C'est le calcul qui suit. Pour simplifier la présentation et les calculs, on suppose C paramétrée par

$$x = t^2 \qquad y = t^3 \qquad z = t^4.$$

L'idéal J de C est alors $(x^3-y^2,\ z-x^2)$ dans $\mathbb{C}[[x,y,z]]$. Le plan $P = \{z = 0\}$ coupe C au voisinage de O suivant l'idéal $I = (z,x^2,y^2)$ qui est celui d'un quadruplet : une base de $\mathcal{O}/I$ est $1,x,y,xy$. <u>Le plan P recoupe C en n-4 autres points supposés simples et distincts</u>.

Si κ est le nombre de cusps de C (tous supposés simples), il y a donc $\kappa(n-4)$ triplets alignés non formés de points simples correspondant à la situation décrite plus haut. On va voir qu'il faut les compter avec <u>multiplicité 3</u>.

Pour faire ce calcul en coordonnées, supposons donc que la droite $\{x=z=0\}$ recoupe C en un point simple $m = (0,1,0)$. Des équations de C au voisinage de m sont alors

$$z = \varphi(y-1) \qquad , \qquad x = \psi(y-1).$$

où $\varphi,\ \psi \in \mathbb{C}[[T]]$ avec $\varphi'(0) \neq 0$ puisque P coupe C transversalement en m. Soit t le triplet formé du doublet d'idéal (y^2,x,z) et du point simple $m = (0,1,0)$. Une carte de $\mathrm{Hilb}^3_c\, \mathbb{P}^3$ en t est fournie par $(a,b,a',b',a'',b'',u,v,w)$ correspondant au doublet d'idéal

$$I = (y^2+ay+b,\ x+a'y+b',\ z+a''y+b'')$$

et au point simple $(u,1+v,w)$. La sous-variété $D \subset \mathrm{Al}^3\, \mathbb{P}^3$ de $\mathrm{Hilb}^3_c\, \mathbb{P}^3$ s'exprime dans cette carte par

$$\begin{cases} (1) \quad u+a'(1+v)+b' = 0 \\ (2) \quad w+a''(1+v)+b'' = 0 \\ (3) \quad a^2-4b = 0 \end{cases}$$

où (1) et (2) expriment $\mathrm{Al}^3\, \mathbb{P}^3$ et (3) la sous-variété D.

Maintenant, toujours dans cette carte, $\mathrm{Hilb}^3 C$ s'exprime par les conditions:

- d'une part : (4) $w = \varphi(v)$, (5) $u = \psi(v)$ concernant le point simple,
- d'autre part, l'inclusion d'idéaux $J = (x^3-y^2, z-x^2) \subset I$ concernant le point double.

Examinons cette dernière condition. Elle est équivalente à l'assertion

$$-(a'y+b')^3 - y^2 \in I \qquad \text{et} \qquad -a''y-b''-(a'y+b')^2 \in I.$$

Autrement dit, ces deux polynômes doivent être multiples de y^2+ay+b.

On obtient les conditions

$$\begin{cases} (6) & 3a'b'^2-a'^3b = a(1+3a'^2b' - a'^3a) \\ (7) & b'^3 = b(1+3a'^2b' - a'^3a) \\ (8) & 2a'b'+a'' = a'^2a \\ (9) & b'^2+b'' = a'^2b \;. \end{cases}$$

C'est un exercice laissé au lecteur de voir que les relations (1) à (9) sont équivalentes à

$$\begin{cases} a' = -(b'+\psi(v))U_1 & w = \varphi(v) \\ v = b'^2U_2 & u = \psi(v) \\ b = a^2/4 & a = 3a'b'^2U_3 \\ \qquad b'^3 = 0 & \\ a'' = 2b'(b'+\psi(v))U_4 & b'' = b'^2U_5 \end{cases}$$

où U_i désigne une unité. Par suite, la longueur de l'idéal considéré est 3. En résumé, on peut énoncer le

THEOREME 2' : Soit C une courbe réduite de degré n dans $\mathbb{P}^3$, ayant h points doubles apparents et pour seules singularités κ cusps simples en position générale (i.e. chaque plan osculateur en un cusp recoupe C en des points simples).

Alors le nombre de tangentes à C recoupant C est

$$n(n-2)(n-5)-2h(n-6)-3\kappa(n-4)$$

correspondant à $k(C)-3\kappa(n-4)$.

On retrouve ainsi la formule de Salmon ([17] , p. 227).

REFERENCES.

[1] A.B. Altman, A. Iarrobino, S.L. Kleiman. Irreducibility of the compactified Jacobian, Proceedings of the Nordic Summer School, Oslo (1976), 1-12.

[2] A. Cayley. On skew surfaces, otherwise scrolls, collected papers V, 168-220.

[3] Enzyklopädie der Mathematischen Wissenschaften, III, 2.2.A.

[4] W. Fulton, R. MacPherson. Intersecting cycles on an algebraic variety, Real and Complex singularities, Oslo 1976, 179-197.

[5] P. Griffiths, J. Harris. Algebraic Geometry, J. Wiley and Sons (1978), New-York.

[6] A. Grothendieck. Les schémas de Hilbert, Séminaire Bourbaki, exposé 221 (1961), IHP, Paris.

[7] A. Grothendieck. EGA I, Publ. Math. IHES 4 (1960), 1-228.

[8] R. Hartshorne. Algebraic Geometry, Springer-Verlag (1977), New-York, Heidelberg, Berlin.

[9] R. Hartshorne. Connectedness of the Hilbert Scheme, Publ. Math. IHES 29 (1966), 261-304.

[10] P. Le Barz. Géométrie énumérative pour les multi-sécantes, Lecture Notes in Mathematics 683 (1977), 116-167.

[11] P. Le Barz. Validité de certaines formules de géométrie énumérative, C.R. Acad. Sc. Paris, 289 (1979), 755-758.

[12] P. Le Barz. Platitude et non platitude de certains sous-schémas de $\mathrm{Hilb}^k \mathbb{P}^N$, à paraître.

[13] P. Le Barz. Quelques calculs dans la variété des alignements, à paraître.

[14] J. G. Semple, L. Roth. Introduction to algebraic geometry, Clarendon Press (1949), Oxford.

[15] F. Severi. Riflessioni intorno ai problemi numerativi... Rend. del R. Ist. Lomb. di Scienze e Lettere, 54 (1921), 243-254.

[16] J. Von zur Gathen. Sekantenraüme von Kurven, thèse Université Zürich (1980).

[17] H. G. Zeuthen. Sur les singularités ordinaires des courbes géométriques à double courbure, C.R. Acad. Sc. Paris, 67 (1868), 225-229.

Patrick LE BARZ
I.M.S.P., Parc Valrose
06034 NICE CEDEX

FRANCE.

SCHUBERT CALCULUS FOR COMPLETE QUADRICS

Israel Vainsencher

1. Introduction. We study the variety parametrizing the complete quadric r-folds in n-space and obtain its Chow ring. Our motivation stems from Kleiman's survey on Hilbert's Problem 15 [4] and the introduction to the reprint of Schubert's Kalkül...[5].

Complete quadric r-folds are defined inductively by the choice of a complete quadric hypersurface in the singular locus of a given quadric r-fold. For instance, complete quadric surfaces (r=2) look like these:

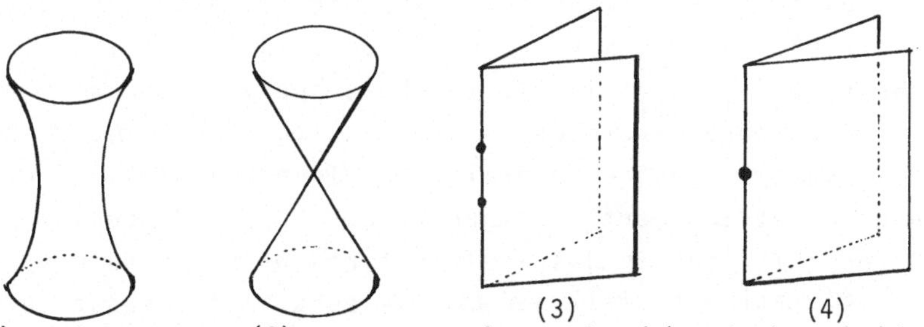

(1) *non degenerate* (2) *cone* (3) (4) *plane pair with 2 foci or double focus*

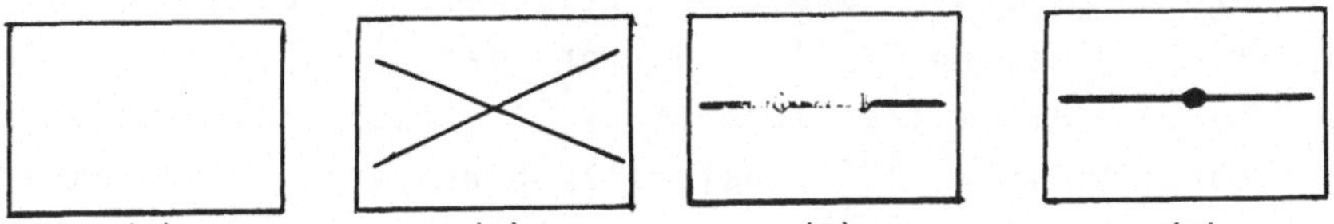

(5) (6) (7) (8)
double plane with complete conic

This was Schubert's ingenious idea for compactifying the variety of non degenerate quadrics in such a manner that the envelopes of tangent lines, planes, etc... be always well defined. E.g., a plane (resp. a line) is taken to be tangent to a plane pair with foci (resp. to a double plane with a conic) iff it hits one focus (resp. the conic). If a plane pair with axis $\underline{a}$ is written as a limit of smooth quadrics varying in a pencil then the envelope of tangent planes of the generic quadric $\underline{q}$ of the pencil specializes to the family of planes through either point of the intersection of $\underline{a}$ and $\underline{q}$. Similar statements valid in arbitrary dimension are also retrieved here (6.10).

The paper is organized as follows. In §2 we review the usual identification "quadric hypersurface" $\longleftrightarrow$ "non zero symmetric map up to scalar multiple". Our starting point is the projective bundle S parametrizing such symmetric maps, along with the determinantal loci $V_r \subseteq S$ parametrizing the family of maps of rank $\leq r$ (3.1). The blowing up of V_r along V_{r-1} is shown to be a projective bundle over a suitable Grassmannian (3.5). In §4 we introduce the sequence e_i of tangent envelope maps. We show that, for a quadric q with associated symmetric map v, the tangent envelope $e_i(q)$ is the quadric associated to an appropriate exterior power of v (4.2). We give here a key formula (4.4) for the normal sheaf of $V_i - V_{i-1}$ in S. In §5 we apply to conics in n-space the machinery developed thus far, with the purpose of setting the pattern for the general case. In particular, we explain how to get Schubert's numbers (such as the 92 conics meeting 8 general lines in 3-space) in terms of Chern numbers (5.3).

§6 contains the main results. We construct a sequence of blowing ups of S along the strict transforms of the V_i's: we put $S^o = S$, $V_i^o = V_i$ and, for $t \geq 1$, let S^t be the blowing up of S^{t-1} along V_t^{t-1} and, for $i \neq t$ (resp. $i = t$), let V_i^t be the strict transform of V_i^{t-1} (resp. the exceptional divisor). We show S^{n-1} naturally parametrizes the complete quadrics in P^n. In fact, these creatures have marvelous geometric properties! Let us sketch here, for simplicity, the case of quadric surfaces in 3-space. Presently S is a projective 9-space wherein V_1 (resp. V_2, V_3) parametrizes the double planes (resp. the plane pairs, the quadric cones). By computing the normal sheaf of V_1, we show that the fibre of V_1^1 over a double plane $q \in V_1$ is the linear system of conics in the plane sing(q). Moreover, the intersections

$V_2^1 \cap V_1^1$ and $V_3^1 \cap V_1^1$ are the families of degenerate conics, namely, double lines and (coplanar) line pairs in 3-space. The next blowing up center, V_2^1, is the P^2-bundle parametrizing the family of plane pairs through a given line. Since choosing a plane through a line is the same as picking a point on the dual line, V_2^1 may also be viewed as the bundle of degree 2 divisors on lines in the dual 3-space. Continuing, we obtain the normal sheaf of V_2^1 in S^1. It reveals that the exceptional divisor V_2^2 is the bundle over V_2^1 of linear systems of degree 2 on lines in 3-space. Deviantly on first sight, V_3^2 is a bundle not of projective spaces, but of varieties of complete conics (P^5 blown up along a Veronese). This is the clue for the general (higher dimensional) case: while $V_i - V_{i-1}$ parametrizes the cones over a non degenerate quadric (i-2)-fold, we prove V_i^{i-1} is fibred over a suitable grassmannian- which accounts for the vertices of the cones- by the variety of complete quadric (i-2)-folds.

The 8 configurations pictured above are each parametrized by a variety of the form

$$\bigcap_{i\varepsilon I} V_i^2 - \bigcup_{j\varepsilon I} V_j^2, \text{ for } I \subseteq \{1,2,3\}.$$

Since we give a complete description of each blowing up center V_i^{i-1}, we may, by general principles, compute the Chow ring of S^i. In particular, we show that Schubert's simple conditions that a complete quadric touch a given subspace of dimension $i=0,\dots,n-1$, form a basis for $A^1(S^{n-1})$. For $n=3$, we describe the basis of $A_1(S^2)$ dual to the latter under intersection pairing. As an application, we retrieve the characteristics of the condition that a quadric surface be tangent to a given surface in terms of its degree, class and rank. Also for $n=3$, we have verified all Schubert's "fundamental numbers" by computing the appropriate Chern numbers (8.8). Unfortunately we did not master the "combinatorics" required to get closed formulas for arbitrary n; A. Lascoux may certainly do if he so desires.

The most delicate step in the study of the V_i^t's rests on a formula relating the Ideals of their strict and total transforms in S^{t+1}. Its proof lead us to a somber incursion through a maze of saturated, homogeneous and normal ideals, better left for an Appendix. The net result is the fact that the Ideal of V_i^{i-1} in $V_j^{i-1} (i<j)$ is normal.

I have the pleasure to thank S. Kleiman for his continued encouragement and to R. Piene for much help during her visit at the Math. Department of UFPE - Recife. I'm indebted to Neide for the typing.

2. Symmetric Maps and Quadrics. We review the usual identification "quadric hypersurface" <—> "non zero symmetric map up to scalar".

Throughout this paper, we denote by X a base scheme over an algebraically closed field k. We denote by E a locally free $\mathcal{O}_X$-Module of rank n+1.

2.1 Let $\mathcal{S}E$ be the subsheaf of $\underline{\mathrm{Hom}}_X(E^*,E)$ of symmetric maps of $\mathcal{O}_X$-Module $v:E^*\to E$, (i.e., $v^*=v$ holds) where * stands for dual. It can easily be seen that $\mathcal{S}E=(\mathrm{sym}_2(E^*))^*$ holds. In particular, $\mathcal{S}E$ is a locally free $\mathcal{O}_X$-Module of rank $\binom{n+2}{2}$. If M is an invertible $\mathcal{O}_X$-Module then the sections of $(\mathcal{S}E)\otimes M$ are the maps $v:E^*\to E\otimes M$ such that $v^*\otimes M=v$; these again are said to be symmetric.

2.2 We denote by S(E) (or simply S) the projective bundle $P(\mathcal{S}E)^*$. It comes equipped with a tautological 1-quotient,

2.2.1 $$(\mathcal{S}E)^*_S \longrightarrow \mathcal{O}_S(1)$$

The corresponding section of $(\mathcal{S}E)\otimes\mathcal{O}_S(1)$ yields the symmetric map.

2.2.2 $$u_E : E^*_S \longrightarrow E\otimes\mathcal{O}_S(1)$$

which is universal for nowhere zero symmetric maps up to scalar, in the following sense. Let T be a scheme /X, let M be an invertible $\mathcal{O}_T$-Module and let $v:E^*_T\to E\otimes M$ be a nowhere zero symmetric map of $\mathcal{O}_T$-Module. Then there exists a unique map $t:T\to S$ of schemes /X such that $v=t^*u_E$.

If char. $k \neq 2$, then $\mathrm{sym}_2(E^*) = (\mathrm{sym}_2 E)^*$ holds. Presently $S(E)$ is the projective bundle $/X$ parametrizing the linear system of quadric hypersurfaces of the fibres of $P(E)$ over X.

3. Degenerate Quadrics.

3.1 Assume, for the moment, char. $k \neq 2$. Let F be a k-vector space of rank $r+1$. Let $v: F^* \to F$ be a nonzero symmetric map and let q denote the associated quadric hypersurface of $P(F)$. Write $\mathrm{sing}(q)$ for the singular locus of q. Then one knows that $\mathrm{sing}(q) = P(\mathrm{coker}(v))$ and $\dim.\mathrm{sing}(q) = r - \mathrm{rank}(v)$. This motivates the study of the determinal loci we now define. Namely, let $V_r = V_r(E)$ denote the scheme of zeros of $\overset{r+1}{\Lambda} u_E$ (2.2.2) in S.

Since $\overset{r+1}{\Lambda} u_E$ is a section of $(s \overset{r+1}{\Lambda} E) \otimes \mathcal{O}_S(r+1)$, the ideal IV_r is the image of the dual of that section,

3.1.1 $$(s \overset{r+1}{\Lambda} E)^* \otimes \mathcal{O}_S(-r-1) \twoheadrightarrow IV_r \subseteq \mathcal{O}_S$$

Clearly, a symmetric map $v \in S$ lies in V_r iff $\mathrm{rank}\,(v) \leq r$. Notice V_o is empty. Also $V_{n+1} = S$. Obviously, V_{r-1} is a closed subscheme of V_r. We state without proof the following easy result.

3.2 Proposition. $V_1(E)$ is equal to the image of the Veronese embedding,

$$P(E^*) \hookrightarrow P(\mathrm{sym}_2(E^*)) = S(E).$$

Moreover, the isomorphism $P(E^*) \cong V_1(E)$ carries the restriction of u_E (2.2.2) to the composition,

$$E^* \twoheadrightarrow L = L^{-1} \otimes L^2 \rightarrowtail E \otimes L^2, \text{ where } L = \mathcal{O}_{P(E^*)}(1).$$ ∎

3.3 Let $G^i = G^i(E)$ be the Grassmann scheme$/X$ with tautological sequence

3.3.1 $$K_i(E) \rightarrowtail E_{G^i} \twoheadrightarrow Q^i(E),$$

where $\mathrm{rank}\; K_i(E) = i$. For simplicity, we will also write

$K_i = K_i(E)$ and $Q^i = Q^i(E)$. We also put $G_i(E) = G^{n-i}(E)$.

3.4 Put $S_i = S_i(E) = S(\overset{i+1}{\Lambda} E)$. Notice we have a closed imbedding $G^i \hookrightarrow S_{i-1}$ via Plücker imbedding $G^i \hookrightarrow P(\overset{i}{\Lambda} E^*)$ followed by Veronese imbedding $P(\overset{i}{\Lambda} E^*) \hookrightarrow S_{i-1}$ (3.2).

We have the following generalization of (3.2).

3.5 <u>Proposition</u>. <u>$S(K_i)$ is equal to the blowing up of V_i along V_{i-1}.</u>

<u>Proof</u>. Set for short $K = K_i$ and $L = \mathcal{O}_{S(K)}(1)$. The diagram

$$\begin{array}{ccc} K^* & \xrightarrow{u_K} & K \otimes L \\ \uparrow & & \downarrow \\ E^* & \longrightarrow & E \otimes L \end{array} \quad (\text{over } S(K)),$$

defines a map $S(K) \to S(E)$. One readily sees that it factors thru $V_i(E)$ and yields the cartesian square

$$\begin{array}{ccc} V_{i-1}(K) & \subset & S(K) \\ \downarrow & & \downarrow \\ V_{i-1}(E) & \subset & V_i(E). \end{array}$$

Observe $V_{i-1}(K)$ is a divisor in $S(K)$ (with invertible Ideal $(\overset{i}{\Lambda} K)^{-2} \otimes L^{-i}$. To complete the proof, let $t: T \to V_i(E)$ be any map such that $J := t^{-1}(IV_{i-1}(E))$ is an invertible Ideal. We must show t factors thru a map $T \to S(K)$. For this, set $M = t^* \mathcal{O}(1)$. Consider the pullback of (3.1.1),

$$w : (s \overset{i}{\Lambda} E)^* \otimes M^{-i} \longrightarrow\!\!\!\!\rightarrow J \subset \mathcal{O}_T$$

Because J is invertible, w defines a map $T \to S_{i-1}$ (3.4), whence a map $t' : T \to V_i(E) \underset{X}{\times} S_{i-1}$. Notice we have $S(K) \hookrightarrow V_i(E) \underset{X}{\times} G^i \subset V_i(E) \times S_{i-1}$ (by 3.4). Since t' maps the open dense subscheme $T - t^{-1}(V_{i-1}(E))$ into $S(K)$, it follows that t' factors thru $S(K)$ (because $S(K)$ is proper $/X$). ■

3.6 Proposition. Assume char. $k \neq 2$. Then we have:

(1) $V_i(E)$ parametrizes the set of quadrics $q \in S(E)$ with singular locus of codimension $\leq i$ in $P(E)$.

(2) $S(K_i)$ is equal to the closure in $S \times G^i$ of the graph of the map defined by $q \in V_i - V_{i-1} \to \mathrm{sing}(q) \in G^i$.

(3) The fibre of $S(K_i)$ over $q \in V_{i-1}$ is the set of all $[n-i]$ lying in $\mathrm{sing}(q)$.

Proof. The assertions follow from (3.1) and (3.5). ∎

4. Tangent Envelopes.

Let $\underline{E} : E_o \rightarrowtail E_1 \twoheadrightarrow E_2$ be an exact sequence of locally free $\mathcal{O}_X$-Modules of ranks m, $n+1$ and $n+1-m$. For each $t=m, m+1,\ldots,n$, there is a natural, surjective map of $\mathcal{O}_X$-Modules,

4.1.1 $$p_t = p_{t,\underline{E}} : (\overset{m}{\Lambda} E_o) \otimes \overset{t+1}{\Lambda} E_1^* \twoheadrightarrow \overset{t-m+1}{\Lambda} E_2^*,$$

defined by the rule

$$p_t(a_1\Lambda\ldots\Lambda a_m \otimes b_o\Lambda\ldots\Lambda b_t)(\bar{c}_o\Lambda\ldots\Lambda\bar{c}_{t-m}) =$$

$$= (b_o\Lambda\ldots\Lambda b_t)(a_1\Lambda\ldots\Lambda a_m\Lambda c_o\Lambda\ldots\Lambda c_{t-m}),$$

for local sections a_j in E_o, b_j in E_1^* and c_j in E_2. (The bar denotes image in E_2).

Taking 2nd symmetric powers of (4.1.1) and recalling (2.1), we get the locally split injection,

4.1.2 $$\varepsilon : \mathcal{S}\, \overset{t-m+1}{\Lambda} E_2 \rightarrowtail \mathcal{S}(\overset{m}{\Lambda} E_o^* \otimes \overset{t+1}{\Lambda} E_1) = (\overset{m}{\Lambda} E_o^*)^2 \otimes \mathcal{S}\, \overset{t+1}{\Lambda} E_1.$$

Now, given a symmetric map v in $S(E_2) - V_{t-m}$, we define its associated imbedded tangent envelope by

$$e_t(v) := (\varepsilon \otimes M^d)(\overset{d}{\Lambda} v)$$

where $d = t-m+1$ and $M = \mathcal{O}_{S(E_2)}(1)$.

Put in other words, $e_t(v)$ fits into the commutative diagram,

$$
\begin{array}{ccc}
\overset{m}{\Lambda} E_0 \otimes \overset{t+1}{\Lambda} E_1^* & \xrightarrow{\ e_t(v)\ } & \overset{m}{\Lambda} E_0^* \otimes \overset{t+1}{\Lambda} E_1 \otimes M^d \\
{\scriptstyle p_t}\big\downarrow & & \big\uparrow {\scriptstyle p_t^* \otimes M^d} \\
\overset{d}{\Lambda} E_2^* & \xrightarrow{\ \overset{d}{\Lambda} v\ } & \overset{d}{\Lambda} (E_2 \otimes M).
\end{array}
\qquad (d=t-m+1) \tag{4.1.3}
$$

The map

$$e_t = e_{t,\underline{E}} : S(E_2) - V_{t-m} \to S_t(E_1) \tag{4.1.4}$$

thus defined is called the <u>(imbedded) tangent envelope map</u> (cf. (4.2) below). We register, for later use, that the pullback of the tautological 1-quotient of $s(\overset{t+1}{\Lambda} E_1)^*$ via e_t fits into the diagram,

$$
\begin{array}{ccc}
(s \overset{t+1}{\Lambda} E_1)^* & \longrightarrow & (s \overset{d}{\Lambda} E_2)^* \otimes L \\
\big\downarrow & & \big\downarrow \\
e_t^* \mathcal{O}_{S_t(E_1)}(1) & = & M^d \otimes L,
\end{array}
\tag{4.1.5}
$$

where $M = \mathcal{O}_{S(E_2)}(1)$ and $L = (\overset{m}{\Lambda} E_0)^{-2}$.

4.2 <u>Proposition</u>. <u>Assume char</u>. $k \neq 2$. <u>Let</u> $\underline{F} : F_0 \to F_1 \to F$ <u>be an exact sequence of</u> k-<u>vector spaces of ranks</u> m, $n+1$ <u>and</u> $n-m+1$. <u>Let</u> $q \in S(F) - V_{t-m}(F)$ <u>have associated symmetric map</u> $v : F^* \to F$. <u>Then the quadric hypersurface</u> $e_t(q) := e_{t,\underline{F}}(v)$ <u>intersects the image (via Plücker imbedding) of the grassmannian</u> $G_t(F_1)$ <u>in</u> $P(\overset{t+1}{\Lambda} F_1)$ <u>in the set of</u> [t]'s <u>touching</u> q.

<u>Proof</u>. Set $d=t-m+1$. Let $h: F \to H$ be a quotient vector space of rank d. The subspace $P(H) \subset P(F)$ touches q iff the composite map $hvh^* : H^* \to H$ is of rank $<d$. Now let $h_1 : F_1 \to H_1$ be a quotient space of rank $t+1$. Then $P(H_1)$ touches q iff its intersection with $P(H)$ contains some [d-1] touching q. Put $g_1 = (\overset{t+1}{\Lambda} h_1) \otimes \overset{m}{\Lambda} F_0$. Set for short

$p = p_{t,\underline{F}}$ (4.1.1). Set $w = p^*.(\overset{d}{\Lambda} v).p$. We will show $P(H_1)$ touches q iff its Plücker coordinates $g_1 \in P(\overset{t+1}{\Lambda} F_1)$ lies in the quadric hypersurface $e_t(q)$, that is, iff $g_1.w.g_1^*=0$. Suppose H is also a quotient of H_1 so that we have a commutative diagram with exact rows,

$$\begin{array}{ccccccc} & F_0 & \rightarrowtail & F_1 & \twoheadrightarrow & F \\ & h_0 \downarrow & & \downarrow h_1 & & \downarrow h \\ \underline{H}\ : & H_0 & \rightarrowtail & H_1 & \twoheadrightarrow & H. \end{array}$$

Put $g_0 = (\overset{m}{\Lambda} h_0) \otimes \overset{t+1}{\Lambda} H_1^*$ and $g = \overset{d}{\Lambda} h$. We construct the diagram

$$\begin{array}{ccccc} \overset{m}{\Lambda} F_0 \otimes \overset{t+1}{\Lambda} H_1^* & \xrightarrow{\ g_0\ } & \overset{m}{\Lambda} H_0 \otimes \overset{t+1}{\Lambda} H_1^* & \xrightarrow{\ f\ } & \overset{d}{\Lambda} H^* \\ & \searrow g_1^* & & g \swarrow & \\ & \overset{m}{\Lambda} F_0 \otimes \overset{t+1}{\Lambda} F_1^* & \xrightarrow{\ p\ } & \overset{d}{\Lambda} F^*, & \end{array} \qquad (d=t-m+1)$$

where f is the isomorphism $p_{t,\underline{H}}$ (4.1.1). Plainly, if $g.(\overset{d}{\Lambda} v).g^* = 0$ then $g_1.w.g_1^* = 0$. Conversely, if the latter holds, we show the former also holds for some h as above. Two cases arise. $1^{\underline{st}}$, if g_0 is injective then the assertion is trivial. Next, if g_0 is not injective then the same is true for h_0. In this case, $P(H_1) \cap P(F)$ contains a subspace of dimension $\geq d$. Since any such $[d]$ contains some $[d-1]$ touching q, we win anyway. ∎

4.3 Remark. For simplicity, we say $e_t(q)$ *is* (or *parametrizes*) the tangent envelope of $[t]$'s touching q, instead of saying it "intersects G_t..." as in the last sentence of (4.2).

It might be worthwhile reading (4.2) in the case of a conic q in 3-space. Thus, take $m=1$, $n=3$. Presently, the quadric $e_1(q)$ is the set of lines meeting q, whereas, if q is not a double line then $e_2(q)$ is the set of planes containing a tangent line to q.

4.4 <u>Proposition.</u> $V_i - V_{i-1}$ <u>is smooth/X and its normal sheaf in</u> S <u>is naturally isomorphic to</u> $(\mathcal{S}Q^i) \otimes \mathcal{O}_S(1)$.

<u>Proof</u>. The 1st assertion, as well as the reason why the 2nd one makes sense is due to the identification $V_i - V_{i-1} = S(K_i) - V_{i-1}(K_i)$ obtained in (3.5). To compute the normal sheaf, we construct the following diagram of maps of $\mathcal{O}_{S(K_i)}$-Modules. We simplify the notation putting $K = K_i(E)$, $M = \mathcal{O}_S(1)$, etc:

4.4.1

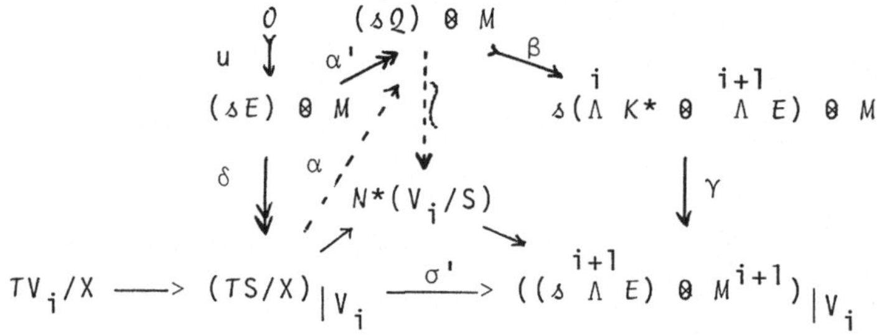

The vertical exact sequence comes from the tautological sequence of $P(\mathcal{S}E)^*$. The map α' takes a local section $E^* \to E \otimes M$ to the composition $Q^* \to E^* \to E \otimes M \to Q \otimes M$. Notice that $\alpha'(u_E) = 0$ because the pullback of u_E to $S(K)$ factors thru u_K (cf. 3.5). Hence α' factors as indicated. Define β by sending a local section $v : Q^* \to Q \otimes M$ to $\beta(v) = (p^* \otimes M).v.p$, where $p = p_{i,\underline{E}}$ (4.4.1) taking $\underline{E}$ to be the tautological sequence of G^i (3.3). Next, define γ starting with the isomorphism $\mathcal{S}(\overset{i}{\Lambda} K^* \otimes \overset{i+1}{\Lambda} E) \otimes M \cong (\mathcal{S}\overset{i+1}{\Lambda} E) \otimes (\Lambda K)^{-2} \otimes M$ and composing it with the map $(\mathcal{S}\overset{i+1}{\Lambda} E) \otimes \overset{i}{\Lambda} K^* \otimes (\overset{i}{\Lambda} u_K) \otimes M$. The bottom part of the diagram is constructed in the lemma below, putting $F = \mathcal{S}E$, $H = \mathcal{S}\overset{i+1}{\Lambda} E$ and $\phi(v_0 \otimes \ldots \otimes v_i) = v_0 \wedge \ldots \wedge v_i$ for local sections v_j in $\mathcal{S}E$. By the same lemma, for each local section v of $(\mathcal{S}E) \otimes M$, we have

$$(\sigma'\delta)(v)(y_0 \wedge \ldots \wedge y_i) = \sum_t u(y_0) \wedge u(y_1) \wedge \ldots \wedge v(y_t) \wedge \ldots \wedge u(y_i)$$

for local sections y_t of E^*. A straightforward computation shows that $\gamma\, \beta\, \alpha' = \sigma'\delta$. Since γ is an isomorphism off

V_{i-1}, it follows that $\gamma\beta$ maps $(\mathcal{S}Q)\otimes M$ isomorphically onto the normal sheaf $N^*(V_i/S)$ over $V_i - V_{i-1}$. ∎

4.5 Lemma. Let F, H be coherent, locally free $\mathcal{O}_X$-Modules. Let $\phi: F^{\otimes m} \to H$ be a $\mathcal{O}_X$-homomorphism. Set $Y = P(F^*)$. Let $w: M^* \to F_{|Y}$ be dual to the universal 1-quotient. Let σ be the section of $H \otimes M^m$ defined by the composition $\phi_Y \cdot w^{\otimes m} : M^{-m} \to H_{|Y}$. Let Z denote the scheme of zeros of σ. Then there is a natural map $\sigma' : (TY/X)_{|Z} \to (H \otimes M^m)_{|Z}$ such that $\ker(\sigma') = TZ/X$ and the composition of σ' with the epimorphism $\delta: F \otimes M \to TY/X$ (restricted to Z) maps local sections $v: M^{-1} \to F$ to $\phi_Y \sum_{j=1}^{m} w\otimes\ldots\otimes v\otimes\ldots\otimes w$.

Proof. The section σ induces the epimorphism $\sigma^* : (H \otimes M^m)^* \twoheadrightarrow I(Z)$. Restricting to Z, we get $(H \otimes M^m)^*_{|Z} \twoheadrightarrow N(Z/Y)$. Composing with the standard exact sequence of conormal and cotangent sheaves of a closed imbedding, then dualizing yields,

$$TZ/X \to \underbrace{(TY/X)_{|Z} \to N^*(Z/Y) \to (H \otimes M^m)_{|Z}}_{\sigma'}.$$

Clearly $\ker(\sigma')$ is equal to TZ/X. The formula for $\sigma'\delta$ follows from an easy, explicit computation, assuming F, H trivial and employing Leibniz's rule for the derivative of a product. ∎

5. Complete Conics. We apply the results gathered thus far to the case of conics in n-space.

5.1 We take $X = \mathrm{Spec}(k)$ and consider the Grassmannian $G = G_2(E)$ with tautological sequence $K = K_{n-2} \rightarrowtail E_G \twoheadrightarrow R$ (rank $R = 3$). If char $k \neq 2$, the P^5-bundle $S(R) \to G$ parametrizes the family of conics of $P^n = P(E)$. We have the map $S(R) - V_1(R) \to S_1(R)$ which assigns to each conic C, not a double line, the envelope of tangent lines lying in the plane of C. Applying (4.1) to the tautological sequence of G, with $m=t=n-2$ (resp. $t=m+1=n-1$), we get imbeddings $S(R) \to G \times S_{n-2}(E))$. (resp. $S_1(R) \to G \times S_{n-1}(E))$. Composing these maps with projection onto the $2^{\underline{nd}}$ factor yields

$S(R) \to S_{n-2}(E)$ (resp. $S(R) - V_1(R) \longrightarrow S_{n-1}(E)$) which assigns to each C the set of $[n-2]$'s meeting C (resp. its envelope of tangent hyperplanes). Let $S^1(R) \to S(R)$ be the blowing up of $S(R)$ along $V_1(R)$. Let $V_1^1(R)$ denote the exceptional divisor. We have the diagram,

5.1.1
$$\begin{array}{ccccc} V_1^1(R) & \hookrightarrow & S^1(R) & \xrightarrow{e} & S_1(R) \\ \downarrow & & \downarrow & \overset{f}{\searrow} & \downarrow \\ V_1(R) & \hookrightarrow & S(R) & & S_{n-1}(E). \end{array}$$

The map e above comes from projection of $S^1(R)$ imbedded in $S(R) \times S_1(R)$; this imbedding is defined by the epimorphism (3.1.1), which now reads,

5.1.2
$$(s \overset{2}{\Lambda} R)^*_{S(R)} \longrightarrow\!\!\!\!\rightarrow IV_1(R) \otimes M^2,$$

where $M = O_{S(R)}(1)$, for the rest of this section.
Since this epimorphism restricts, off $V_1(R)$, to the epimorphism defining the tangent line envelope correspondence ((4.1.4) with $t=1$, $m=0$) it follows that $S^1(R)$ is equal to the closure of the graph of that correspondence. Pulling back (5.1.2) to $S^1(R)$, we find the commutative diagram,

5.1.3
$$\begin{array}{ccc} (s \overset{2}{\Lambda} R)^*_{S^1(R)} & \longrightarrow\!\!\!\!\rightarrow & IV_1^1(R) \otimes M^2 \\ \uparrow\!\!\uparrow & \nearrow\!\!\!\nearrow & \\ (s(\overset{n-2}{\Lambda} K^* \otimes \overset{n}{\Lambda} E))^*_{S^1(R)} & & \end{array}$$

where the vertical arrow comes from (4.1.2). The slant arrow defines f (5.1.1). Thus, the pullback of the tautological 1-quotient $O_{S_{n-1}(E)}(1)$ via f is equal to $IV_1^1(R) \otimes M^2 \otimes (\overset{n-2}{\Lambda} K)^{-2}$.
Recalling the formula for the conormal sheaf N of V_1 in S, namely, $N = (sQ^1)^* \otimes M^*$ (4.4) where $Q^1 = Q^1(R)$ we get

$$V_1^1(R) = P(N) = S(Q^1)$$

and

$$\mathcal{O}_{P(N)}(1) = (IV_1^1)_{V_1^1} = \mathcal{O}_{S(Q^1)}(1) \otimes \mathcal{O}_{S(R)}(-1)$$

Presently, $P(Q^1)$ is the P^1-bundle/$V_1(R)$ whose fibre over each double line $C \in V_1(R)$ is the line sing C. Also, $S(Q^1)$ parametrizes the positive divisors of degree 2 on the fibres of $P(Q^1)$ over $V_1(R)$. Hence, each double line C gives rise to the ∞^2 many choices of "pairs of foci" on sing C. This explains Fig. (7).

Pulling back (5.1.2) to V_1^1 yields the diagram,

$$\begin{array}{ccc}
(s \overset{2}{\Lambda} R)^*|_{V_1^1} & \longrightarrow & \mathcal{O}_{S(Q^1)}(1) \otimes M \\
\downarrow & \searrow & \uparrow \\
(N \otimes M^2)|_{V_1^1} & = & (sQ^1)^* \otimes M,
\end{array}$$

where the slant arrow is $(\gamma\beta)^* \otimes M^2$ (from 4.4.1, with $i=1$ and E replaced by R and Q replaced by $Q^1(R)$). Presently the map γ is an isomorphism because V_0 is empty. By construction of β (cf. 4.2 and 4.4.1), it follows that the restriction of f to $V_1^1(=SQ^1)$ maps each double line with pair of foci to the point in $S_{n-1}(E)$ representing the hyperplane pair of $P(\overset{n}{\Lambda} E) = P(E^*)$ which parametrizes the family of hyperplanes of $P(E)$ containing either focus. In particular, a pair (C,C') in $S(R) \times S_{n-1}(E)$, such that C is a double line, belongs to $S^1(R)$ iff C' is the hyperplane pair dual to a pair of points on C. Now it is easy to see that the action on $S^1(R)$ induced from the general linear group of E has precisely the 4 orbits $V_1(Q^1)$, $V_1^1(R) - V_1(Q^1)$, $V_2(R) - V_1(R)$ and $S(R) - V_2(R)$.

5.2 The Chow ring of $S^1(R)$ can easily be described because $S^1(R) \to S(R)$ is the blowing up of the projective bundle $S(R) \to G$ along the subvariety $V_1(R) \cong P(R^*)$ with normal sheaf $N^* = M \otimes sQ^1(R)$. For instance, for $n \geq 3$, the piece of codim. 1 cycles is freely generated by the classes $\mu := c_1R$, $\theta := c_1M$ and $\lambda = [V_1^1]$. (If $n=2$ then $E=R$ and θ, λ suffice). Now let ν (resp. ρ) be the pullback of the hyperplane class of $S_{n-2}(E)$ (resp. $S_{n-1}(E)$). By construction of the maps from $S^1(R)$ to $S_{n-2}(E)$ and to $S_{n-1}(E)$ (cf. 5.1.3) we get

$$\nu = \theta + 2\mu \quad \text{and} \quad \rho = \lambda + 2\theta + 2\mu.$$

Furthermore, ν (resp. ρ) is the class of the condition that a conic meet a given [n-2] (resp. be tangent to a [n-1]). Thus, the classes μ, ν, ρ also form a basis of $A^1(S^1(R))$.

The basis of $A_1(S^1(R))$ (= cycles of dim.1) dual to μ, ν, ρ under intersection pairing consists of the classes μ', ν', ρ' defined by the pencils pictured below:

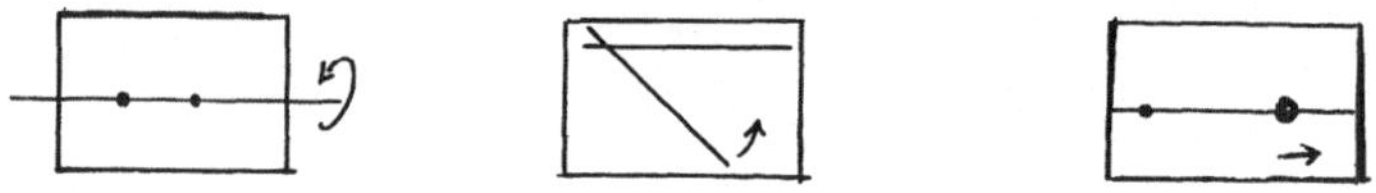

μ' (move the plane) $\qquad$ ν' (move one line) $\qquad$ ρ' (move one focus)

It follows that any simple condition $D \in A^1(S^1(R))$ may be written as $D = \alpha\mu + \beta\nu + \gamma\rho$, where $\alpha = D.\mu'$, $\beta = D.\nu'$ and $\gamma = D.\rho'$. For instance, take $n=3$ and let Y be an integral surface in P^3 such that the dual map of a general plane section is birational. Let d, d' denote degree and rank of Y. Let D be the closure in $S^1(R)$ of the set of smooth conics tangent to Y at a smooth point. One checks that a complete conic lying in $V_1^1(R)$ is in D iff its line is tangent to Y or Y contains one of the foci. Thus, $D.\mu' = 0$. It can also be shown that a general pencil defining ν' (resp. ρ') intersects D transversaly. Hence get $D.\nu' = d'$ and $D.\rho' = d$. Therefore, $D = d\rho + d'\nu$ holds in $A^1(S^1(R))$.

5.3 The computation of intersection numbers $\mu^a \nu^b \rho^c$, where $a+b+c = \dim S(R)$, is now a purely mechanical matter. One uses the basis change relations (5.2.1) and apply projection formula to the maps induced on the Chow rings by the maps in (5.1.1). The main ingredient needed here are the Chern classes of $\mathcal{S}R$ and $\mathcal{S}Q^1(R)$. For example, for $n=3$ we may use the exact sequence

$$\underline{\mathrm{Hom}}(K^*,E) = K \otimes E \rightarrowtail \mathcal{S}E \twoheadrightarrow \mathcal{S}R.$$

We find, for the total Chern class,

$$c(\mathcal{S}R) = c(K \otimes E)^{-1} = (1-\mu)^{-4}$$

$$c(\mathcal{S}Q^1(R)) = c(\mathcal{S}R)\ c(K_1(R) \otimes R)^{-1}.$$

Using these formulas, we have verified all numbers found by Schubert, e.g. $\nu^8=92$ (conics meeting 8 general lines in 3-space) and $\rho^8=4$ (conics tangent to 8 general planes).

6. <u>Complete Quadric r-Folds in n-Space</u>. We study the closure of the graph of the tangent envelope maps. Consider the Grassmannian $G=G_{r+1}(E)$ with tautological sequence $K \rightarrowtail E_G \twoheadrightarrow R$, with rank $R = r+2$. We have the "intrinsic" tangent envelope maps ${}_ie\colon S(R) - V_i(R) \to S_i(R)$ defined by ${}_ie(v) = \Lambda^{i+1} v$ for $v\colon R^* \to R \otimes M$ of rank $\geq$ i+1 (i=0,...,r). Now set t=i+n-r-1; we have also define "imbedded" tangent envelope maps $e_t\colon S(R) - V_i(R) \to S_t(E)$ in (4.1.4) (taking $\underline{E}$ to be the tautological sequence of G and m=n-r-1). There are imbeddings $S_i(R) \hookrightarrow G \times S_t(E)$, defined by (4.1.2) so that e_t is just the composition of ${}_ie$ with projection. Consequently, the closure of the graph of ${}_1e\times\ldots\times{}_ie$ in $S(R) \times S_1(R)\times\ldots\times S_i(R)$ is equal to that of $e_{n-r}\times\ldots\times e_t$ in $G \times S_{n-r-1}(E)\times\ldots\times S_t(E)$. For this reason, we may as well restrict to the case r=n-1, so that "imbedded" and intrinsic" are one and the same.

6.1 Put $S^0=S$ and $V_i^0=V_i$. For $t\geq 1$, let $S^t \to S^{t-1}$ be the blowing up of V_t^{t-1}; then, for $t\neq i$ (resp. t=i) let V_i^t denote the strict transform of V_i^{t-1} (resp. the exceptional divisor). We write also $S^t=S^t(E)$, etc..., when necessary.

6.2 We define a <u>complete symmetric map</u> (csm) on E to be a flag $E = E_0 \twoheadrightarrow E_1 \twoheadrightarrow \ldots \twoheadrightarrow E_m$ of locally free quotient Modules together with a choice of a point v_t in $S(E_t)$ such that v_m is bijective and $E_t^* = \ker(v_{t-1})$ for t=1,...,m.

The sequence $r_t = n+1-\text{rank } E_t$, with $1 \leq t \leq m$, is called the <u>type</u> of the csm. The empty sequence is allowed for $m=0$; it is the type of bijective symmetric maps. We assume $r_m \leq n$ for $m \geq 1$. Thus, there are 2^n possible types. Notice that rank $(v_t) = r_{t+1} - r_t$.

6.3 <u>Theorem</u>. (1) <u>For each</u> $t=1,\ldots,n-1$, <u>there is a natural imbedding of</u> S^t <u>into</u> $S \times S_1 \times \ldots \times S_t$ <u>which maps</u> S^t <u>onto the closure of the graph of the tangent envelope maps</u> $e_1 \times \ldots \times e_t : v \longrightarrow (\overset{2}{\Lambda} v, \ldots, \overset{t+1}{\Lambda} v)$;

(2) S^t <u>is smooth over</u> X;

(3) <u>For each (possibly empty) sequence</u> $\rho = (r_1 < \ldots < r_m)$, <u>between</u> 1 <u>and</u> n, <u>the scheme</u>

$$V_\rho := \bigcap_{i \in \rho} V_i^{n-1} - \bigcup_{t \notin \rho} V_t^{n-1}$$

<u>is smooth</u>/X <u>and parametrizes the csm of type</u> ρ. <u>If</u> $X=\text{Spec}(k)$ <u>then the</u> V_ρ <u>are precisely the orbits of the action induced by the general linear group of</u> E <u>on</u> S^{n-1}.

<u>Proof</u>. We argue by induction on $n = \text{rank } E - 1$. We need a few auxiliary results which are of independent interest. Before stating them in full generality, we explain in details the first steps as motivation.

We get from (3.1.1), with $r=1$, that $S^1 = \text{Proj}(\oplus (IV_1)^m)$ imbeds in $S \times S_1 = S \times P(s\overset{2}{\Lambda} E)^*$. Clearly, the image of the open dense subset $S - V_1$ is equal to the graph of $e_1 : S - V_1 \to S_1$. Since V_1 and S are smooth/X, so is S^1. Now recall that the conormal sheaf of V_1 in S is naturally isomorphic to $(sQ^1)^* \otimes O_S(-1)$ (4.4). Therefore, we get $V_1^1 = S(Q^1)$. Next, by (3.5), the blowing up V_2^1 of V_2 along V_1 is equal to $S(K_2)$. Thus, V_2^1 is smooth/X, whence so is S^2. The main step ahead is the following

6.3.1 <u>Claim</u>: $\qquad (IV_2)O_{S^1} = (IV_2^1).(IV_1^1)^2.$

Assuming this for the moment, we get (by 3.1.1), the epimorphism

$$(s\overset{3}{\Lambda}E)^* \otimes \mathcal{O}_S(-3) \otimes (IV_1^1)^{-2} \twoheadrightarrow IV_2^1 .$$

Therefore, $S^2 = \mathrm{Proj}(\bigoplus (IV_2^1)^m)$ imbeds in $S^1 \times S_2$ which already lies in $S \times S_1 \times S_2$. Since $S^2 - V_1^2 - V_2^2$ maps onto the graph of $e_1 \times e_2$, we see that S^2 is indeed the closure of that graph. We further remark that we have $V_2^1 \cap V_1^1 = V_1(\mathcal{Q}^1)$ in $V_1^1 = S(\mathcal{Q}^1)$. Indeed, we clearly have $V_2^1 \cap V_1^1 = V_1(K_2)$ in $V_2^1 = S(K_2)$ (cf. proof of 3.5). The final identification $V_1(\mathcal{Q}^1) = V_1(K_2)$ follows from (3.2) via the natural identification $G^1(K_2) = G^1(\mathcal{Q}^1)$ explained in the diagram below:

$$\begin{array}{ccccc}
K_1(K_2) & \rightarrowtail & K_2 & \twoheadrightarrow & \mathcal{Q}^1(K_2) = K_1(\mathcal{Q}^1) \\
\| & & \downarrow & & \downarrow \\
K_1 & \rightarrowtail & E & \twoheadrightarrow & \mathcal{Q}^1 \\
 & & \downarrow & & \downarrow \\
 & & \mathcal{Q}^2 & = & \mathcal{Q}^1(\mathcal{Q}^1).
\end{array}$$

To prove the claim (6.3.1) we need the following.

6.4 Lemma. *Let R be a local ring. Let I, J be ideals of R. Let $f_1, \ldots, f_t$ be elements of J. Suppose the initial forms of the f_i in $gr_I(R) = \bigoplus I^P/I^{P+1}$ generate $gr_I(J,R) = \bigoplus J \cap I^P + I^{P+1}/I^{P+1}$. Let R' be a monoidal transform of R with center I and let x in I satisfy $IR' = xR'$. Set $m_i = \max\{m \mid f_i \in I^m\}$. Then the strict transform J' of J in R' is generated by $f_1 x^{-m_1}, \ldots, f_t x^{-m_t}$. In particular, if all $m_i = m$ then $JR' = x^m.J'$.*

Proof. See Hironaka (/1/p.216) or Valla and Valabrega (/9/Cor. 2.4). ■

Proof of (6.3.1). We apply the lemma to a local ring R of $S(E)$. For the sake of simplicity, we assume X smooth, connected. Thus, R is a regular local ring. As V_1, V_2 are integral schemes, their respective ideals I, J in R are prime. Furthermore, $gr_I(R)$ is the polynomial ring $\mathrm{sym}(I/I^2)$. Now $\mathrm{Proj}(gr_I(R))$ is equal to the exceptional divisor $(V_1^1)|_R$

of the blowing up of Spec(R) along I. Set $\bar{R} = R/J$, $\bar{I} = I/J$. Then $\mathrm{Proj}(gr_{\bar{I}}(\bar{R}))$ is the exceptional divisor $(V_1^1 \cap V_2^1)_{|R}$ of the blowing up of $\mathrm{Spec}(\bar{R})$ along $\bar{I}$. Recalling the exact sequence

$$gr_I(J,R) \hookrightarrow gr_I(R) \twoheadrightarrow gr_{\bar{I}}(\bar{R}),$$

we see that $gr_I(J,R)$ is a homogeneous ideal defining the closed subscheme $(V_1^1 \cap V_2^1)_{|R}$ in $(V_1^1)_{|R}$. On the other hand, since $V_1^1 \cap V_2^1 = V_1(Q^1)$ in $S(Q^1)$, there is another obvious homogeneous ideal defining $V_1(Q^1)_{|R}$, namely, the prime ideal H generated by the 2×2 minors of a symmetric matrix whose entries are indeterminates over R/I forming a basis for I/I^2. Since $gr_I(J,R)$ is saturated (cf. Appendix, Cor. 5), it follows that $gr_I(J,R)$ is equal to H, hence is generated by elements of degree 2 (i.e., lying in I^2/I^3). These elements are initial forms of elements of J. By the lemma, it follows that, for each local ring R' of S^1 dominating R, the strict transform J' of J satisfies $JR' = x^2J'$ where $xR' = IR'$. This proves (6.3.1).■

Continuing with the proof of (6.3) we register that the equality $V_2^1 \cap V_1^1 = V_1(Q^1)$ in $V_1^1 = S(Q^1)$ implies $V_1^2 = S^1(Q^1)$. The next step is the following

6.5 Assertion. $V_3^2 = S^1(K_3)$.

To prove this, we construct the diagram of maps,

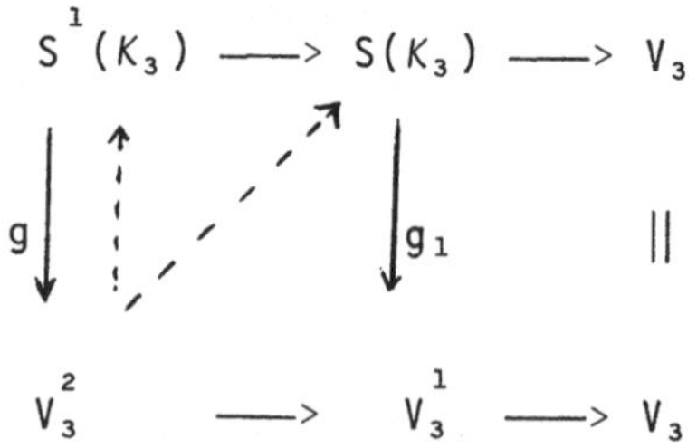

where the horizontal maps are blowing ups. The map g_1 exists because $(IV_1)\mathcal{O}_{S^1(K_3)} = IV_1(K_3)\mathcal{O}_{S^1(K_3)} = IV_1^1(K_3)$ is invertible. To construct g, notice $(IV_2^1)\mathcal{O}_{S^1(K_3)} = (IV_2)(IV_1^1)^{-2}\mathcal{O}_{S^1(K_3)} =$

$= IV_2(K_3)(IV_1^1(K_3))^{-2} = IV_2^1(K_3)$ which is invertible. A similar computation gives the map $V_3^2 \to S(K_3)$ and finally $V_3^2 \to S^1(K_3)$ inverse to g. ∎

Next, we claim:

6.6 <u>The conormal sheaf</u> N_2 <u>of</u> V_2^1 <u>in</u> S^1 <u>is naturally isomorphic to</u> $(sQ^2)^* \otimes \mathcal{O}_S(-1) \otimes I(V_1^1)^{-1}{}_{|V_2^1}$.

<u>Proof of (6.6)</u>. Set for short $M:=\mathcal{O}_S(1)$, $I: = IV_1^1$ and $N': = (sQ^2)^* \otimes M^{-1}$. Construct the following diagram of cotangent maps over V_2^1,

$$\begin{array}{ccccccc}
N' & \dashrightarrow & IN_2 & \to & N_2 & \to & N_2/IN_2 = N_1 \\
 & {\scriptstyle \alpha^*}\searrow & \cap\,\vdots & & \downarrow & & \downarrow \\
 & & \Omega_{S/X} & \xrightarrow{\phi} & \Omega_{S^1/X} & \overset{\psi}{\twoheadrightarrow} \Omega_{S^1/S} = & \Omega_{V_1^1/V_1} \\
 & & & & \downarrow{\scriptstyle \eta} & & \downarrow \\
 & & & & \Omega_{V_2^1/X} & \to \Omega_{V_2^1/V_2} = & \Omega_{V_2 \cap V_1^1/V_1}.
\end{array}$$

The vertical sequences starting at N_2 and N_1 are the usual exact sequences for the regular imbeddings of V_2^1 in S^1 and $V_2 \cap V_1^1$ in V_1^1 (the latter one being extended by zero). The formula $N_1 = N_2/IN_2$ for the conormal sheaf of $V_2^1 \cap V_1^1$ in V_1^1 holds because that intersection is transversal. The horizontal exact sequence ϕ, ψ is restriction to V_2^1 of the usual sequence for a map $S^1 \to S$ of schemes/X. The map ϕ is injective because it is an isomorphism off V_1^1 and $\Omega_{S/X}$ is locally free. The split injection α^* is dual to α in (4.4). By (4.4), $\eta\phi\alpha^*$ vanishes off V_1^1. Since $\Omega_{V_2^1/X}$ is locally free, therefore $\eta\phi\alpha^*$ vanishes on V_2^1. Thus $\phi\alpha^*$ factors through N_2. Since the diagram commutes, if follows that $\phi\alpha^*$ actually factors through $I.N_2$. Since

$I.\Omega_{S^1/X}$ lies in $(\Omega_{S/X})_{S^1}$ we finally get the inclusion $IN_2 \hookrightarrow \Omega_{S/X}$. Dualizing[1] the latter, we get the commutative triangle,

$$\begin{array}{ccc} \Omega^*_{S/X} & & \\ {\scriptstyle\alpha}\downarrow & \searrow & \\ N'^* & \longleftarrow & I^{-1}\, N_2^* \end{array}$$

where the bottom arrow is injective because it is so off V_1^1 and N_2 is locally free. Therefore $N_2 = I^{-1} \otimes N'$ as asserted. ∎

The formula for N_2 yields at once the identifications,

$$V_2^2 = S(Q^2) \times_{G^2} S(K_2)$$

and

6.6.1 $$O_{S(Q^2)}(1)\,|_{V_2^2} = (IV_1^2).(IV_2^2)\; O_{S(E)}(1)\,|_{V_2^2}\,.$$

We now state formally the steps required to complete the induction.

6.7 Theorem. Let X be a smooth, connected variety. Let E be a locally free O_X-Module of rank $n+1$. Then, with the notation of (6.1) and (3.3), the following hold for all $m<n$:

(1,m) $V_t^{t-1} = S^{t-1}(K_t)(=S^{t-2}(K_t)$ if $t\geq 2)$ for $t<m$;

(2,m) there is an isomorphism of the normal sheaf N_t^* of V_t^{t-1} in S^{t-1} with

$(IV_{t-1}^{t-1})(IV_{t-2}^{t-1})\dots(IV_1^{t-1})\,(\mathfrak{s}Q^t) \otimes O_S(1)$ (restricted to V_t^{t-1}) which restricts to (4.4) off $\bigcup_{i<t} V_i$, for $t\leq m-1$;

---

[1] *I'm indebted to R. Piene for this trick.*

(3,m) $V_t^{t+r} = S^r(\mathcal{Q}^t) \times_{G^t} S^{t-1}(K_t)$, <u>for</u> $1 \le t+r \le m-1$;

(4,m) $V_t^{t-1} \cap V_r^{t-1} = V_r^{t-2}(K_t)$, <u>for</u> $1 \le r < t \le m$;

(5,m) $V_t^{t-1} \cap V_r^{t-1} = V_{t-r}^{t-r-1}(\mathcal{Q}^r) \times_{G^r} S^{r-1}(K_r)$, <u>for</u> $1 \le r < t \le m$;

(6,m) $(IV_i^{t-1})\mathcal{O}_{S^t} = (IV_i^t)(IV_t^t)^{i-t+1}$, <u>for</u> $1 \le t < i \le m$.

<u>Proof</u>. We have already dealt with the cases $n=2$ and 3. Now assume the theorem for locally free Modules of rank less than $n+1$. We prove assertions (1,m) through (6,m) by induction on m. We have already checked the cases $m=1$ and 2.

(1,m+1): Put $t=m+1$. We must prove $V_t^{t-1} = S^{t-2}(K_t)$. (The equality $S^{t-1}(K_t) = S^{t-2}(K_t)$ holds because $V_{t-1}^{t-2}(K_t)$ is a hypersurface). As in the proof of (6.5), we construct the diagram,

$$\begin{array}{ccccccccc}
S^{t-2}(K_t) & \xrightarrow{f_{t-2}} & --- & \xrightarrow{f_2} & S^1(K_t) & \xrightarrow{f_1} & S(K_t) & \xrightarrow{f} & V_t \\
\downarrow g & \searrow^{g_{t-2}} & & & & \searrow^{g_1} & & & \| \\
V_t^{t-1} & \xrightarrow[h_{t-2}]{} & V_t^{t-2} & \longrightarrow & --- & \xrightarrow[h_1]{} & V_t^1 & \xrightarrow[h]{} & V_t
\end{array}$$

where the horizontal maps are successive blowing ups. There exists a (unique) map g_1 such that $g_1^{-1}((IV_1^1)\mathcal{O}_{V_t^1}) = IV_1^1(K_t)$. This follows from the universal property of the blowing up h because $f_1^{-1}f^{-1}((IV_1)\mathcal{O}_{V_t}) = IV_1^1(K_t)$ is invertible. Similarly, using (6,m) (with m=2), we may write

$$f_2^{-1}g_1^{-1}((IV_2^1)\mathcal{O}_{V_t^1}) = f_2^{-1}g_1^{-1}((IV_1^1)^{-2}(IV_2)\mathcal{O}_{V_t^1}) =$$

$$= f_2^{-1}((IV_1^1(K_t))^{-2}.f_1^{-1}(IV_2(K_t)) = f_2^{-1}(IV_2^1(K_t)) = IV_2^2(K_t),$$

which is invertible, thereby defining a map g_2 such that $g_2^{-1}((IV_2^1)\mathcal{O}_{V_t^1}) = IV_2^2(K_t)$. We continue this way up to g_{t-2}. At this last step we get, by repeated use of (6,m),

6.7.1 $(IV_m)O_{S^{m-1}} = (IV_1^{m-1})^m.(IV_2^{m-1})^{m-1}.\dots.(IV_{m-1}^{m-1})^2(IV_m^{m-1}).$

The same formula holds with K_{m+1} in place of E. Since $IV_m(K_{m+1})$ is invertible, therefore g_{t-2} factors through g. As in the proof of (6.5), one shows g is an isomorphism. This proves (1,m+1). ∎

(2,m+1): We compute the conormal sheaf $N_m = N_m(E)$ of V_m^{m-1} in S^{m-1}. Put $N' = (\mathcal{S}Q^m)^* \otimes O_S(-1)$. Consider the diagram of cotangent maps,

$$\begin{array}{ccccc} N' & \xrightarrow{\alpha^*} & \Omega_{S/X} & & \\ \downarrow & & \downarrow{\scriptstyle\phi} & & \\ N_m & \rightarrowtail & \Omega_{S^{m-1}/X} & \xrightarrow{\eta} & \Omega_{V_m^{m-1}/X} \end{array} \qquad \text{(over } V_m^{m-1}\text{).}$$

As in the proof of (6.6), α^* is dual to α in (4.4.1) and $\eta\phi\alpha^*$ vanishes. Thus $\phi\alpha^*$ factors through N_m. We show next that $\phi\alpha^*$ actually factors through $I_r N_m$, where $I_r := IV_r^{m-1}$ for $r=1,\dots,m-1$. By (4,m), we have,

$$V_m^{m-1} \cap V_r^{m-1} = V_r^{m-2}(K_m).$$

Studying the commutative diagram (cf. proof of (6.6)),

$$\begin{array}{ccccccc} N' & \dashrightarrow & I_r N_m & \hookrightarrow & N_m & \twoheadrightarrow & N_m/I_r N_m = N(V_m^{m-1} \cap V_r^{m-1}/V_r^{m-1}) \\ \downarrow & & & & & & \downarrow \\ \Omega_{S^{r-1}/X} & \hookrightarrow & \Omega_{S^r/X} & \twoheadrightarrow & \Omega_{V_r^r/V_r^{r-1}} & \nearrow & \Omega_{V_r^{m-1}/V_r^{r-1}}, \end{array}$$

we obtain $N' \subseteq I_r N_m$. Since N_m is locally free and the intersection $V_1^{m-1} \cap \dots \cap V_m^{m-1}$ is transversal (by 4,m), we have

$$\cap(I_r.N_m) = (\cap I_r).N_m = (\Pi I_r).N_m.$$

Since $(\Pi I_r).\Omega_{S^{m-1}/X}$ lies in (the pull back of) $\Omega_{S/X}$, we get $(\Pi I_r).N_m$ inside $\Omega_{S/X}$ and we may argue as in the proof of (6.6) to get $N' = (\Pi I_r).N_m$ as desired. ∎

(3,m+1) : For $t=m$ and $r=0$, we must show

$$V^m_m = S(Q^m) \underset{G^m}{\times} S^{m-1}(K_m).$$

This follows from (1,m) and (2,m+1). For $t<m$, it remains to show

$$V^m_t = S^{m-t}(Q^t) \underset{G^t}{\times} S^{t-1}(K_t).$$

Now V^m_t is the blowing up of V^{m-1}_t along $V^{m-1}_t \cap V^{m-1}_m$. By (5,m), the latter intersection is just

$$V^{m-t-1}_{m-t}(Q^t) \underset{G^t}{\times} S^{t-1}(K_t).$$

Since

$$V^{m-1}_t = S^{m-t-1}(Q^t) \underset{G^t}{\times} S^{t-1}(K_t)$$

(by 3,m) and since S^r is the blowing up of S^{r-1} along V^{r-1}_r, the assertion is proven. ∎

(4,m+1): We must show $V^m_{m+1} \cap V^m_r = V^{m-1}_r(K_{m+1})$ (under the identification $V^m_{m+1} = S^{m-1}(K_{m+1})$ (1,m+1)), for $r=1,\ldots,m$. We have established, in the course of the proof of (1,m+1), that $S^{r-1}(K_{m+1})$ maps to V^{r-1}_{m+1} and

$$(IV^{r-1}_r) \,.\, O_{S^{r-1}(K_{m+1})} = IV^{r-1}_r(K_{m+1}) \qquad (r=1,\ldots,m).$$

This implies

$$(IV^m_r)O_{S^{m-1}(K_{m+1})} = IV^{m-1}_r(K_{m+1}),$$

which is just the ideal-theoretic version of the desired equality.

(5,m+1): By (4,m+1), we have, for $r=1,\ldots,m$,

$$V^m_{m+1} \cap V^m_r = V^{m-1}_r(K_{m+1})$$

which is equal to

$$S^{m-r-1}(\mathcal{Q}^r(K_{m+1})) \underset{G^r(K_{m+1})}{\times} S^{r-1}(K_r(K_{m+1})),$$

by (3,m).

Since $G^r(K_{m+1}) = G^{m+1-r}(\mathcal{Q}^r)$ over G^r, with $\mathcal{Q}^r(K_{m+1}) = K_{m+1-r}(\mathcal{Q}^r)$ and $K_r(K_{m+1}) = K_r(E)$, we get

$$V^m_{m+1} \cap V^m_r = S^{m-r-1}(K_{m+1-r}(\mathcal{Q}^r)) \underset{G^r}{\times} S^{r-1}(K_r)$$

(by 1,m+1-r) $$= V^{m-r}_{m+1-r}(\mathcal{Q}^r) \underset{G^r}{\times} S^{r-1}(K_r),$$

thus proving (5,m+1). ∎

(6,m+1) We must show,

$$(IV^{t-1}_{m+1})\mathcal{O}_{S^t} = (IV^t_{m+1})(IV^t_t)^{2+m-t} \quad (1 \le t \le m).$$

We observe $V^t_{m+1} \cap V^t_t$ is the image of $V^m_{m+1} \cap V^m_t$ via the structure map $S^m \to S^t$. The latter intersection being equal to

$$V^{m-t}_{m+1-t}(\mathcal{Q}^t) \underset{G^t}{\times} S^{t-1}(K_t),$$

inside

$$V^m_t = S^{m-t}(\mathcal{Q}^t) \underset{G^t}{\times} S^{t-1}(K_t),$$

(by (5,m+1)), we get

$$V^t_{m+1} \cap V^t_t = V_{m+1-t}(\mathcal{Q}^t) \underset{G^t}{\times} S^{t-1}(K_t),$$

inside

$$V^t_t = S(\mathcal{Q}^t) \underset{G^t}{\times} S^{t-1}(K_t).$$

The rest of the argument now goes as in the proof of (6.3). Thus, we apply again the lemma (6.4), now taking R to be a local ring of $S^{t-1}(E)$, letting I and J be the ideals of

V^{t-1} and V^{t-1}_{m+1} in $\mathbb{R}$. We have $\mathrm{Proj}(gr_I(R)) = (V^t_t)_{|R}$. Set $\bar{R} = R/J$, $\bar{I} = I/J$. We have $\mathrm{Proj}\ gr_{\bar{I}}(\bar{R}) = (V^t_t \cap V^t_{m+1})_{|R}$. The prime homogeneous ideal H of the latter in $gr_I(R)$ is generated by the minors of order $2+m-t$ of a symmetric matrix of indeterminates over R/I that form a basis of I/I^2. Since $gr_I(J,R)$ is saturated (cf. Appendix) and defines the same closed subscheme of $(V^t_t)_{|R}$ as H does, we see that $H = gr_I(J,R)$ is generated by elements of degree $2+m-t$ as required. ∎

Having complete the proof of (6.7), it is clear that (1) and (2) of (6.2) are also proven.

6.8 To prove the remaining assertion (3) of (6.7), we add the following formulas to the list in (6.7):

(7,n) $\quad IV^{t+r}_t = (IV^{t-1}_t)\mathcal{O}_{S^{t+r}}$ $\quad$ <u>for all</u> t,r;

(8,n) $\quad V^{t+r+i}_{t+r} \cap V^{t+r+i}_t = V^{r+i}_r(\mathcal{Q}^t) \underset{G^t}{\times} S^{t-1}(K_t)$, $\quad$ } <u>for all</u> t, r, i.

(9,n) $\quad V^{t-1+i}_{t-r} \cap V^{t-1+i}_t = S^{i-1}(\mathcal{Q}^t) \underset{G^t}{\times} V^{t-1}_{t-r}(K_t)$.

The 1st of these follows from the fact that the total and the strict transforms of a hypersurface coincide provided it intersects transversally the center of the blowing up. The 2nd one follows from the 1st together with (5,n). The 3rd follows from (3,n) and (4,n).

Repeated application of (8,n) (with $t+r+i=n-1$) shows that $\bigcap_{i\in\rho} V^{n-1}_i$ is smooth.

To see that V_ρ parametrizes the csm of type ρ, consider first the simple case when the sequence ρ reduces to a single term, say $\rho = (r)$, with $1 \leq r \leq n$. We have

$$V^{n-1}_{r+i} \cap V^{n-1}_r = V^{n-r-1}_i(\mathcal{Q}^r) \underset{G^r}{\times} S^{r-1}(K_r)$$

and

$$V^{n-1}_{r-i} \cap V^{n-1}_r = S^{n-r-1}(\mathcal{Q}^r) \underset{G^r}{\times} V^{r-1}_{r-i}(K_r).$$

Therefore, we get

$$V_\rho = V_r^{n-1} - \bigcup_{j\neq r} V_j^{n-1}$$

$$= (S-V_{n-r})(Q^r) \times_{G^r} (S-V_{r-1})(K_r).$$

Now it should be clear that to give a point in V_ρ is the same as to give an exact sequence of locally free sheaves $H \rightarrowtail E \twoheadrightarrow E_1$ $(\mathrm{rank}(H)=r)$ together with bijective maps $v: H^* \longrightarrow H \otimes L$, $v_1: E_1^* \longrightarrow E_1 \otimes M$. This yields a map $v_o: E^* \twoheadrightarrow H^* \xrightarrow{v} H \otimes L \rightarrowtail E \otimes L$ such that $\ker(v_o) = E_1^*$, and conversely, any csm (v_o, v_1) of type ρ as above gives a point of V_ρ.

For the general case, given $\rho = (r_1 < r_2 < \ldots < r_m)$, set $r=r_1$ and define $\rho' = (r_2-r<\ldots<r_m-r)$. Now we have

$$V_\rho = \bigcap_{i>1} (V_r^{n-1} \cap V_{r_i}^{n-1}) - \bigcup_{\substack{j\notin\rho \\ j>r}} (V_r^{n-1} \cap V_j^{n-1}) - \bigcup_{j<r} V_r^{n-1} \cap V_j^{n-1}$$

$$= (\bigcap_{i\in\rho'} V_i^{n-r-1}(Q^r) - \bigcup_{i\notin\rho'} V_i^{n-r-1}(Q^r)) \times_{G^r} (S-V_{r-1})(K_r)$$

$$= V_{\rho'}(Q^r) \times_{G^r} (S-V_{r-1})(K_r).$$

By induction, one sees readily that a point of V_ρ consists of the following data: (1) an exact sequence of locally free sheaves $H \rightarrowtail E \twoheadrightarrow E_1$ $(\mathrm{rank}\, H=r)$; (2) a csm of type ρ' on E_1 and (3) a bijective symmetric map $v: H^* \longrightarrow H \otimes L$. Just as above, v induces a map $v_o: E^* \longrightarrow E \otimes L$ with $\ker(v_o) = E_1$, giving rise to a csm of type ρ on E. Conversely it is clear that any csm of type ρ defines a unique point of V_ρ as asserted.

The proof of the assertion about the action of the general linear group $G\ell(E)$ on $S^{n-1}(E)$ is easy and will be omitted. ■

Remark. W. Ihle [3] gives a different proof for the smoothness of the closure of the graph in case $X = \mathrm{Spec}\,\mathbb{C}$. See also Tyrrel [8].

7. Tangency to Complete Quadrics. The next proposition tells us how the tangent envelope map $S - V_i \to S_i$ extends to a map $S^i \to S_i$. The geometrical interpretation is given in the corollary.

7.1 Proposition. The following diagram commutes,

$$\begin{array}{ccc} S^r(Q^i) \times_{G^i} S^{i-1}(K) = V_i^{i+r} & \hookrightarrow & S^{i+r} \\ \swarrow & & \downarrow f \\ S^r(Q^i) \xrightarrow{g} S_r(Q^i) \overset{h}{\hookrightarrow} G^i \times S_{i+r} & \longrightarrow & S_{i+r}, \end{array}$$

where f,g are defined by projection of $S^t \subset S \times \ldots \times S_t$ and the imbedding h comes from (4.1.2) (with E equal to the tautological sequence of G^i and $i=m$ and $r=t-m$).

Proof. We take for simplicity $r=0$. The surjection $(s \overset{i+1}{\Lambda} E)^* \otimes O_S(-i-1) \twoheadrightarrow IV_i$ (3.1.1) yields, up on S^{i-1}, a map onto $(IV_i)O_{S^{i-1}} = (IV_1^{i-1})^i \ldots (IV_{i-1}^{i-1})^2\, IV_i^{i-1}$ (by 6.7). Restriction to V_i^{i-1} gives,

$$\begin{array}{c} (s \overset{i+1}{\Lambda} E)^* \twoheadrightarrow N_i \otimes O_S(i+1) \otimes ((IV_1^{i-1})^i \ldots (IV_{i-1}^{i-1})^2)_{|V_i^{i-1}} \\ \| \\ (sQ^i)^* \otimes O_S(i) \otimes ((IV_1^{i-1})^{i-1} \ldots (IV(_{i-1}^{i-1}))_{|V_i^{i-1}} \end{array}$$

(by 6.7). Now recall $V_i^{i-1} = S^{i-1}(K_i)$ and $V_t^{i-1} \cap V_i^{i-1} = V_t^{i-1}(K_i)$ for $t<i$. Since

$$(IV_{i-1}(K_i))O_{S^{i-1}(K_i)} = (IV_1^{i-1}(K_i))^{i-1} \ldots (IV_{i-1}^{i-1}(K_i)) = (\overset{i}{\Lambda} K_i)^{-2} O_S(-i)$$

we finally get the surjection,

$$(s \overset{i+1}{\Lambda} E)^* \twoheadrightarrow (sQ^i)^* \otimes (\overset{i}{\Lambda} K_i)^{-2}.$$

Its restriction over $V_i - V_{i-1}$ is dual to (4.1.2) (with $t=m=i$). Therefore, the resulting map

$$V_i^i = (SQ^i)_{|V_i^{i-1}} \to S_i(E)_{|V_i^{i-1}},$$

composed with projection $S_i(E)_{V_i^{i-1}} \to S_i(E)$ factors as stated. ∎

6.10 Corollary. Assume char.k≠2. Let $q \in V_i - V_{i-1}$ be written as a limit of a 1-parameter family of non degenerate quadrics. Then there exists a quadric hypersurface q' in sing(q) such that the limit of the tangent envelopes is the quadric parametrizing the [i]'s meeting q'.

Proof. The assertion follows from the Proposition above together with (4.2). ∎

7. Basis of A^1 and A_1.

7.1 Theorem. Let $e^i : S^{n-1} \to S^i$ be the (extended tangent envelope map) induced by projection of $S^{n-1} \hookrightarrow S_o \times \ldots \times S_{n-1}$ to S_i. Let $\underline{m}_i$ denote the pullback of the hyperplane class of S_i via e^i. Let $\underline{d}_i$ denote the class of V_i^{n-1} in $A^1(S^{n-1})$. Then each of $\{\underline{m}_o, \ldots, \underline{m}_{n-1}\}$ and $\{\underline{m}_o, \underline{d}_1, \ldots, \underline{d}_{n-1}\}$ is a basis for $A^1(S^{n-1})/A^1(X)$. Moreover, we have the change of basis relations,

$$\underline{m}_t = (t+1)\underline{m}_o - \sum_1^t (t+1-i)\underline{d}_i \qquad (t=1,\ldots,n-1)$$

and

$$2c_1(E) + (n+1)\underline{m}_o = \sum_1^n (n+1-i)\underline{d}_i .$$

Proof. Fix $t \geq 1$. Let ${}^te : {}^tS \to S$ be the blowing up along V_t. By (6.7.1) (with $m=t$) and formula (7,n) (p.25); we have

$$(IV_t)O_{S^{n-1}} = (IV_1^{n-1})^t \ldots (IV_{t-1}^{n-1})^2 (IV_t^{n-1}).$$

Therefore, the structure map $e^o : S^{n-1} \to S$ factors thru ${}^te$. We get the commutative diagram,

$$\begin{array}{ccccc} S^{n-1} & & \xrightarrow{\quad e^t \quad} & & S_t \\ {\scriptstyle e^0}\downarrow & \searrow & & & \\ & & {}^tS & \longrightarrow & S_t. \\ S & \xleftarrow{\ t_e\ } & & & \end{array}$$

Since the pull back of the hyperplane class of S_t to ${}^tS$ is $c_1((IV_t)\mathcal{O}_S(t+1)\mathcal{O}_{{}^tS})$, and since $c_1(IV_i^{n-1}) = -\underline{d}_i$, the first formulas follow. Finally, since $IV_n = (\overset{n+1}{\Lambda} E)^{-2} \otimes \mathcal{O}_S(-n-1)$, the last formula is also proven. Since $A^1(S^i)$ is equal to $A^1(S^{i-1}) \oplus \mathbb{Z}\cdot[V_i^i]$ for $i \geq 1$ and $A^1(S) = A^1(X) + \mathbb{Z}\, c_1\mathcal{O}_S(1)$, the proof is complete. ■

7.2 Assume henceforth $X = \mathrm{Spec}(k)$, char. $k \neq 2$. Presently, a hyperplane of S_i may be taken as the set of all quadric hypersurfaces in $P(\overset{i+1}{\Lambda} E)$ passing through a fixed point which we choose to be (Plücker coordinates of) a point in G_i. Hence $\underline{m}_i$ is the condition that a (complete!) quadric hypersurface of $P^n = P(E)$ be tangent to a fixed $[i]$.

7.3 Let $L_{o,n}$ be the pencil in S^{n-1} defined by fixing a flag $[n-2] \subset [n-1]$ and a complete quadric hypersurface in $[n-2]$ and letting a variable $[n-1]$ move around $[n-2]$. For $i \geq 1$, define $L_{i,n}$ by fixing a double $[n-1]$ and choosing a $L_{i-1,n-1}$ in it. Denote by $\underline{m}_i'$ the class of $L_{i,n}$ in $A_1(S^{n-1})$. Here are the pictures for $n=3$:

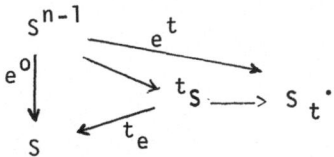

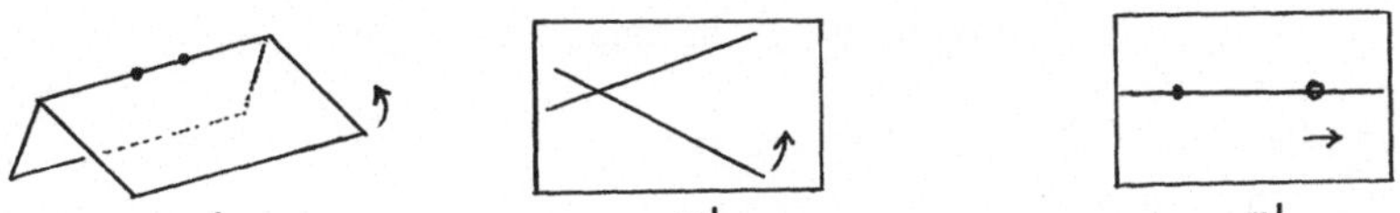

$\underline{m}_o'$ (move one plane) $\underline{m}_1'$ (move one line) $\underline{m}_2'$ (move one point)

7.4 <u>Proposition.</u> $\{\underline{m}_o', \ldots, \underline{m}_{n-1}'\}$ <u>is the basis of</u> $A_1(S^{n-1})$ <u>dual to</u> $\{\underline{m}_o, \ldots, \underline{m}_{n-1}\}$ <u>under intersection pairing.</u>

Proof. We take $n=3$ for simplicity. We must show $\underline{m}_i \bullet \underline{m}'_j = \delta_{ij}$. Since $\underline{m}'_o$ is the class of the strict transform of a line in S, we have $e^o_\bullet(\underline{m}'_o)$ = class of a line. By projection formula, $e^o_\bullet(\underline{m}_o . \underline{m}'_o) = (c_1 \mathcal{O}_S(1)) . e^o_\bullet(\underline{m}'_o) = 1$. Next, $\underline{m}'_o . \underline{m}_1 = 0$ because the image of $L_{o,3}$ under e^1 is the point of S_1 representing the double hyperplane of lines that meet the axis of $L_{o,3}$. Similarly, $\underline{m}'_i . \underline{m}_j = 0$ for all $i \neq j$. To compute $\underline{m}'_1 . \underline{m}_1$, notice $L_{1,3}$ is the proper transform of a pencil in V^1_1 (lying in a fibre over $G^1 = V_1$). Since $V^2_1 \to S_1$ factors thru $V^1_1 \to S_1$, and since the pullback of a hyperplane class of S_1 induces hyperplane class in the fibres of $V^1_1 \to G^1$, we get $\underline{m}'_1 . \underline{m}_1 = 1$. ∎

7.5 Corollary. For any simple condition $D \in A^1(S^{n-1})$ we have,

$$D = \Sigma\, (D . \underline{m}'_i) \underline{m}_i . \quad ∎$$

7.6 As an illustration, we apply this formula to the condition of tangency to a surface in P^3. Let Y be an integral surface in P^3. Let Y' (resp. Y'') denote the closure in $Y \times P^{3*}$ (resp. $Y \times S$) of the set of pairs (y,h) (resp. (y,q)) such that Y is smooth at y and h(resp. q) is a plane (resp. smooth quadric) tangent to Y at y. Let Y^* (resp. Y) be the image of Y' (resp. Y'') in P^{3*} (resp. S). Let D denote the strict transform of Y in S^2. Finally, set

$$Y = \{(y,h,q) \in Y' \times S^2 \mid y \in e^o(q),\ h \in e^2(q) \text{ and each line of } h \text{ containing } y \text{ belongs to } e^1(q)\}.$$

We claim that D is the image of Y in S^2. Indeed, Y is irreducible because all fibres of $Y \to Y'$ are irreducible of equal dimension 6. Since $Y \to S^2$ is generically finite, it follows that the dimension of its image is 8. Since the image clearly contains an open dense subset of D, it is equal to D as we claimed.

Now one may easily prove that:

(1) A plane pair h_1, h_2 with foci y_1, y_2 lies in D iff some (y, h_i) is in Y' or some y_i is in Y;

(2) A double plane h with a line pair ℓ_1, ℓ_2 lies in D iff (y,h) is in Y' for some y or there is some (y,h') in Y' such that ℓ_1 or ℓ_2 contains y and lies in h';

(3) A double plane h with a double line ℓ and foci y_1, y_2 lies in D iff (y,h) is in Y' for some y or else for some (y,h') in Y' we have $y \in \ell \subset h'$ or Y contains y_1 or y_2.

7.7 Proposition. Assume the dual maps of Y and of a general plane section are birational. Let d, d' and d'' denote the degree, rank and class of Y. Then we have $D = d''\underline{m}_0 + d'\underline{m}_1 + d\underline{m}_2$ in $A^1(S^2)$.

Proof. The assumption on Y implies there are pencils $L_{i,2} (i=0,1,2)$ intersecting D transversally. On the other hand, by (1), (2) and (3) above it is clear that, "set theoretically", we have $\underline{m}'_0 . D = d''$, $\underline{m}'_1 . D = d'$ and $\underline{m}'_2 . D = d$. The assertion now follows from (7.5). ∎

8. Computations. Let us say a few words on how to proceed for the calculation of intersection numbers $\underline{m}_0^{a_0} . \underline{m}_1^{a_1} \ldots \underline{m}_{n-1}^{a_{n-1}}$, with $\Sigma a_i = \dim S$. First, one uses the change of basis relations to replace the $\underline{m}_i (i \geq 1)$ by the classes of the "degenerate forms" $\underline{d}_i$. Then one uses the intersection tables such as (8,n) (p.)together with projection formula. Finally, one uses the Chern classes of the conormal sheaves N_i computed in (6.7).

For instance, take $n=3$. We consider the quadric surfaces in 3-space. We use Schubert's notation, setting

$$\mu = \underline{m}_0, \ \nu = \underline{m}_1, \ \rho = \underline{m}_2$$

and

$$\phi = \underline{d}_1, \ \psi = \underline{d}_2, \ \chi = \underline{d}_3 .$$

The change of basis relations (7.1) give,

8.1 $$\phi = 2\mu-\nu, \quad \psi = -\mu+2\nu-\rho \quad \text{and} \quad \chi = -\nu+\rho.$$

One checks that ν (resp. ρ) is the class of the strict transform W_1^2 (resp. W_2^2) of the quadric (resp. cubic) hypersurface W_1 (resp. W_2) of S parametrizing the quadrics of P^3 tangent to a given line (resp. plane). The smooth locus of W_i consists of quadric surfaces simply tangent to the given [i]. Moreover, the tangent hyperplane to W_i at such a smooth point q is the set of all quadrics passing thru the (unique!) point of tangency of q and [i]. Since W_i^2 intersects properly each of the 8 orbits, it follows that the intersection number $\mu^a\nu^b\rho^c$ $(a+b+c=9)$ is the exact number of nondegenerate quadrics passing thru a general points and tangent to b general lines and to c general planes.

Using (8.1), Schubert reduced the computation of $\mu^a\nu^b\rho^c$ $(a+b+c=9)$ to the computation of $x\mu^a\nu^b\rho^c$ for $x = \phi,\psi,\chi$ and $a+b+c=8$.

8.2 To compute $\psi\mu^a\nu^b\rho^c$, construct the diagram,

$$\begin{array}{ccccccc} S(\mathcal{Q}) \times_G S(K) = V_2^2 & \xrightarrow{f} & V_2^1 = S(K) & \xrightarrow{g} & G \\ \big\downarrow i & & & & \\ S^2 & & & & \end{array}$$

where $G = G^2$, etc... Now set

$$\lambda = c_1(\mathcal{Q}), \quad \pi = c_2(\mathcal{Q}) \quad \text{and} \quad \xi = c_1\, \mathcal{O}_{S(\mathcal{Q})}(1).$$

For simplicity, we often drop the pullbacks and write $\lambda = g^\cdot\lambda = f^\cdot g^\cdot\lambda$, etc. We have the formulas,

$$\mu = c_1\, \mathcal{O}_{S(K)}(1)$$

$$i^\cdot\phi = [V_2^2 \cap V_1^2] = f^\cdot[V_1(K)] = f^\cdot c_1(\Lambda^2 K \otimes \mathcal{O}_{S(K)}(1))$$

$$= 2(\mu-\lambda),$$

$$i^{\bullet}\psi = i^{\bullet}(-c_1 IV_2^2) = \mu-\xi-\phi \quad \text{(by 6.6.1)}.$$

By (8.1) we get,

$$i^{\bullet}\nu = 2\lambda \quad \text{and} \quad i^{\bullet}\rho = \xi+2\lambda.$$

By projection formula and invariance of degree of a zero cycle under pushing forward, we find, for $a+b+c = 8$.

$$\begin{aligned}\psi\mu^a\nu^b\rho^c &= i^{\bullet}\mu^a\nu^b\rho^c \\ &= g_{\bullet}f_{\bullet}i^{\bullet}(\mu^a\nu^b\rho^c) \\ &= (2\lambda)^b g_{\bullet}(\mu^a f_{\bullet}(i^{\bullet}\rho^c)).\end{aligned}$$

We list the formulas

$$c(\mathcal{Q}) = 1+\lambda+\pi \qquad \text{(Chern classes of } \mathcal{Q});$$

$$\lambda^4 = 2,\ \pi^2 = \lambda^2\pi = 1;$$

$$\sum_0^4 f_{\bullet}(\xi^{2+r}) = c(\mathcal{S}\mathcal{Q})^{-1} = 1-3\lambda+7\lambda^2-4\pi+\ldots$$

$$\sum_0^4 f_{\bullet}(i^{\bullet}\rho^{2+r}) = 1+3\lambda+7\lambda^2-4\pi+\ldots;$$

$$\sum_0^4 g_{\bullet}(\mu^{2+r}) = c(\mathcal{S}K)^{-1} = 1+3\lambda+3\lambda^2+4\pi+20\lambda\pi-5\lambda^3+16\pi^2-21\lambda^4+36\lambda^2\pi.$$

Using these, we checked all values for $\psi\mu^a\nu^b\rho^c$ found by Schubert ([7] p. 104) are right. Thus, we get zeros unless $2\leqq a,c\leqq 6$ and $b\leqq 4$ by dimension reasons. We have, for example,

$$\psi\mu^2\nu\rho^5 = \psi\mu^5\nu\rho^2 = (20\lambda\pi-5\lambda^3)(2\lambda) = 40-20=20, \text{ etc...}$$

The computations for ϕ and χ in place of ψ are similar and we 've checked they agree with Schubert's. It would be nice to find closed formular for $\Pi \underline{m}_i^{a_i}$ and $\Pi \underline{d}_i^{a_i}$ for the general case.

APPENDIX

NORMAL IDEALS

1. Satureted ideals. Let $A = \bigoplus_{0}^{\infty} A_p$ be a graded ring. Set $A_+ = \bigoplus_{1}^{\infty} A_p$, the so called irrelevant ideal. The saturation of a homogeneous ideal $H = \bigoplus H_p$ in A is the ideal

$$H' = \bigcup_r \{x \in A \mid x.A_+^r \subseteq H\}.$$

H is said to be saturated if $H=H'$. If A_1 generates A then one can easily see that H is saturated iff $(H_{p+r}:A_r) \subseteq H$ for all p,r.

2. Proposition. Let $\mathcal{O}$ be a ring and let I,J be ideals in $\mathcal{O}$. Assume:

(1) I contains a nonzero divisor mod. J

(2) $I + J/J$ is normal in $\mathcal{O}/J$.
Then $gr_I(J,\mathcal{O}) = \bigoplus I^p \cap J + I^{p+1}/I^{p+1}$ is saturated.

Proof. Since $gr_I(\mathcal{O})$ is generated by I/I^2, it suffices to check that

$$(*) \qquad (J+I^{p+r} : I^r) \subseteq J+I^p \quad \text{for all } p,r.$$

For this, let (x_i) be a (finite!) set of generators of I^r. Now if x lies in $(J+I^{p+r} : I^r)$, we get equations

$$x \cdot x_i = y_i + \Sigma z_{ij}x_j, \quad \text{with} \quad z_{ij} \in I^p \text{ and } y_i \in J.$$

By Cramer's rule, we get

$$\det(x\delta_{ij}-z_{ij})x_t \in J \quad \text{for all } t.$$

Since each I^r also contains a nonzero divisor mod. J, it follows that $\det(x\delta_{ij}-z_{ij}) \in J$. Thus, we find a relation,

$$x^d+z_1x^{d-1}+\ldots+z_d = 0 \quad (\text{mod. } J)$$

where $z_i \in (I^p)^i$. Since, by definition of normality, I^p is integrally closed mod. J, we finally get $x+J$ lies in I^p+J/J, thus proving (*).■

3. Proposition. Let be given the blowing up diagram,

$$\begin{array}{ccc} Y' & \overset{j}{\hookrightarrow} & X' \\ f\downarrow & & \downarrow g \\ Y & \hookrightarrow & X. \end{array}$$

Assume that:

(1) X, X' are normal, integral schemes and Y a reduced, nowhere dense, closed subscheme of X;

(2) There exists a projective bundle $h:P \to Y$ and a closed imbedding $i:Y' \to P$ such that $hi=f$ and $i^*(\mathcal{O}_P(1)) = j^* \mathcal{O}_{X'}(1)$ $(=j^*\mathcal{I}(Y'))$;

(3) f is flat, $R^1 f(\mathcal{O}_{Y'}(p)) = 0$ and the natural map $h_*(\mathcal{O}_P(p)) \to f_*(\mathcal{O}_{Y'}(p))$ is surjective for $p \geq 0$; or

(3') f is flat of rel. dim. ≥ 2 and for each geometric point y in Y the fibre Y'_y is regular in codim. 1 and arithmetically Cohen-Macaulay in P_y.

Then the Ideal $\mathcal{I}(Y)$ is normal.

Proof. Since X,X' are normal, $g_*(\mathcal{O}_{X'}(p))$ is the integral closure of $I(Y)^p$ (by Lemma on p. 354 of Zariski-Samuel [10]). Thus, it suffices to show $\mathcal{I}(Y)^p = g_*(\mathcal{O}_{X'}(p))$ for all $p \geq 0$. We adapt an argument of Hironaka ([2], p. 14). Accordingly, it is enough to show:

(*): conditions (1), (2) and (3') imply (3).

For this, we may replace Y by Spec(k). Let H be a general hyperplane of P. Since Y' is R_1, its affine cone Y also is R_1. Hence Y is normal by Serre's criterion. Applying the same argument to $H\cap Y'$, we see that both Y' and $Y'\cap H$

are projectively normal in P. Now construct the diagram,

$$\begin{array}{ccccccc} h_*(\mathcal{O}_P(p)) & \twoheadrightarrow & h_*(\mathcal{O}_H(p)) & & & & \\ \downarrow\!\!\downarrow & & \downarrow\!\!\downarrow & & & & \\ f_*(\mathcal{O}_{Y'}(p)) & \longrightarrow & f_*(\mathcal{O}_{Y'\cap H}(p)) & \longrightarrow & R^1f(\mathcal{O}_{Y'}(p-1)) & \longrightarrow & R^1f(\mathcal{O}_{Y'}(p)). \end{array}$$

By projective normality, the vertical maps are surjective. Since $R^1f(\mathcal{O}_{Y'}(p)) = 0$ for $p>>0$, the assertion (*) is proven. ∎

4. Corollary. The Ideal of V_t^{t-1} in V_m^{t-1} is normal for all $t<m\leq n$.

Proof. Apply 3. to $Y: = V_t^{t-1}$ and $X:=V_m^{t-1}$. Now X and X' $(=V_m^t)$ are normal schemes by Serre's criterion. On the other hand, we have

$$Y' = V_t^t \cap V_m^t = V_{m-t}(Q^t)_Y$$

in $P : = S(Q^t)$. Thus, Y' is flat/Y with geometric fibres as in (3') of the above Proposition for $n-t\geq 2$. (Cf. Jozefiak [4] or Kutz [6] for the Cohen-Mecaulayness assumption). Finally, for $t=n-1 = m-1$, we have $Y' = V_1(Q^{n-1})_Y = P(Q^{n-1})^*_Y$ imbedded in P via Veronese and (3) applies. ∎

5. Corollary. Let I and J be the Ideals of V_t^{t-1} and V_m^{t-1} in S^{t-1} $(t<m\leq n)$. Then $gr_I(J,\mathcal{O}_{S^{t-1}})$ is saturated in $gr_I(\mathcal{O}_{S^{t-1}})$.

Proof. The assertion follows from Prop. 2 and Cor. 4. ∎

REFERENCES

[1] H. Hironaka, "Resolution of singularities...", Annals of Math. 79 (1964) 109-326.

[2] H. Hironaka, "Smoothing of algebraic cycles..." Am. J. Math. 90 (1968) 1-54.

[3] W. Ihle, "Untersuchungen über vollständige Quadriken", Inauguraldissertation, Martin-Luther Univ., Halle-Wittenberg (1968).

[4] T. Józefiak, "Ideals generated by minors of a symmetric matrix", Comment. Math. Helv. 53 (1978), 595-607.

[5] S. Kleiman, "Problem 15. Rigorous foundation of Schubert enumetative calculus", Proceeding of Symp. P. Math, 28, A.M.S., Providence (1976).

[6] R. E. Kutz, "Cohen-Macaulay rings..." Trans. A. M. S. 194 (1974) 115-129.

[7] H. Schubert, "Kalkül der abzählenden Geometrie" reprint, Springer-Verlag (1979).

[8] J. A. Tyrrell, "Complete quadrics and collineations in S_r", Mathematika 3 (1956), 69-79.

[9] G. Valla & P. Valabrega, "Standard bases and generators for the strict transforms", preprint Istituto Mat. Torino, n. 15 (1979).

[10] O. Zariski and P. Samuel, "Commutative Algebra" vol. II.

Departamento de Matemática
Universidade Federal de Pernambuco
Cidade Universitária
Recife - Pe. Brasil.

MULTIPLE POINT FORMULAS FOR MAPS[1]

Steven L. Kleiman[2]

M.I.T. 2-278

Cambridge, MA 02139

U.S.A.

Let $f: X \to Y$ be a map. Its set of r-fold points is

$$M_r = \{x \varepsilon X | \text{ there exist } x_2, \ldots, x_r \text{ with } f(x_i) = f(x)\};$$

the x_i must be distinct from x and from each other, but they may lie "infinitely close" (that is, determine tangent directions along the fiber $f^{-1}f(x)$). An r-fold-point formula is a polynomial expression in the invariants of f which gives, under appropriate hypotheses, the number of r-fold points, weighted by natural multiplicities, or the class m_r of a natural positive cycle supported by M_r. The theory of these formulas will be surveyed here, concentrating on some of the author's recent work, Kleiman [1981b], [1982]. Aside from a few comments, the setting will be algebraic geometry, although the formulas and their proofs have a universal character.

One of the oldest and most familiar multiple-point formulas is due to Clebsch (1864). It deals with a smooth complete curve X and an immersion f of X into a plane Y such that the image $Z = f(X)$ has nodes but no worse singularities. The formula is

$$\varepsilon_0 = 1/2(\mu_0 - 1)(\mu_0 - 2) - p \qquad \text{or} \tag{1}$$

$$2\varepsilon_0 = \mu_0{}^2 - (3\mu_0 - (2-2p)) \tag{2}$$

where ε_0 is the number of nodes of Z and μ_0 is its degree and p is the genus of X.

Every node of Z is the image of 2 points of X, each of which is a double-point of the immersion $f: X \to Y$. So $2\varepsilon_0$ is equal to the number of points of M_2, which in turn is equal to the degree of the linear-equivalence class m_2 of the divisorial cycle $[M_2]$ associated to M_2. Therefore Formula (2) can be obtained by extracting degrees, term by term, from the following refined formula:

$$m_2 = f^* f_* m_1 - c_1 m_1 , \tag{3}$$

where m_1 denotes the (fundamental) class of X and c_1 is given by the formula,

$$c_1 = f^* c_1(Y) - c_1(X). \tag{4}$$

Put another way, c_1 is the Euler class of f, the "top" Chern class of the virtual normal bundle of f.

Formula (3) is part of the classical theory of adjoints (initiated by Riemann in 1857), for it says just that, for any curve W in Y of degree (μ_0-3) that passes through the nodes of Z, the (positive) cycle $f^* [W] - [M_2]$ is the cycle of a (holomorphic) differential. Today, the theory of adjoints is subsumed under general Grothendieck duality theory, which correspondingly yields a generalization of (3) as follows.

Let X and Y be smooth, projective algebraic varieties or, more generally, flat, locally projective, finitely presentable S-schemes with Gorenstein fibers. Let $f: X \to Y$ be a finite map such that $Z = f(X)$ is a divisor in Y flat over S. By general Grothendieck duality theory (see, for example, Kleiman [1980], (18), (19)), we have

$$\omega_{X/S} = C \otimes f^* \omega_{Y/S} \otimes f^* O_Y(Z) \tag{5}$$

where C is the "conductor" on X of X/Z,

$$C = \underline{\mathrm{Hom}}_Z (f_* O_X, O_Z)^{\sim} . \tag{6}$$

Formula (5) embodies the essence of the theory of adjoints. Now, suppose that f carries X birationally onto Z. Then C endows M_2 with the structure of an effective divisor; let m_2 denote the corresponding linear-equivalence class. With (4) in mind, set

$$c_1 = f^* c_1 (\omega_{Y/S} {}^*) - c_1 (\omega_{X/S} {}^*). \tag{7}$$

Then (5) yields a formula, which looks just like (3) and generalizes it.

The preceding approach of adjoints (which was first presented in Kleiman [1977] Ch. V, Sect. A) is very different in spirit from the general approach, based on the residual-intersection theorem, which is the subject of most of the remainder of the article. Fulton [1978] Thm. 3 checked that, when X is a smooth curve and Y is a smooth surface over an algebraically closed field, the two approaches yield the same structure of divisor on M_2, and essentially the same thing was done independently by Fischer, Gussein-Zade and Tessier, according to Fulton [1978] Rem. 2.6. However, it remains unproved that the two structures agree for more general X and Y. It is to be hoped that future work will provide a greater link between the approaches.

Around the last half of the 19th century, geometers sought and found relations among various numerical characters of m-dimensional algebraic loci in projective (m + n)-space. In particular, the enumeration of the double-points culminated in the following lovely formula of Severi (1902):

$$2\varepsilon_0 = \mu_0(\mu_0 - 1) - \sum_{j=1}^{n} \mu_j , \tag{8}$$

where ε_0 is the degree (order) of the double locus and μ_j is the degree of the locus of points whose tangent m-plane meets a general codimension-(m + j)-plane. (This is the so-called "polar" locus, and it may also be thought of as the ramification locus of a general central projection onto an (m + j - 1)-space.) Severi's derivation of (8), which is based on the correspondence principle, was recently brought up to today's standards by Catanese [1979].

A version of Severi's formula (8) for the case m = n was established by Peters and Simonis (1974) using the Grassmannian of lines. The general formula is implicit in Holme's work (1974) on embedding obstructions. Further work along the same lines was done in 1975-76 by Holme, Holme-Roberts, Johnson, and Laksov. All this work is surveyed in Kleiman [1977], Ch. V, Sect. B. It led in a spate of activity in 1976 to modern multiple-point theory.

The first general double-point formula was found by Todd (1940); he obtained it in rational (linear) equivalence along with a residual-intersection formula, one formula from the other by induction on the dimension. B. Segre (1953) included Todd's formulas in his theory of covariants of immersion. Independently, Whitney (1941) gave a double-point formula for an immersion of differentiable manifolds. Ronga [1973], inspired by Whitney, obtained the double-point formula in ordinary cohomology for a generic map with ramification. However, it was Laksov [1978a], inspired to some extent by Ronga, who recognized that, while Todd's reasoning is specious, nevertheless the residual-intersection formula could be derived on its own and used to prove the double-point formula.

Let $f: X \to Y$ be a map of smooth, projective varieties over an algebraically closed field, and consider the diagram,

$$\begin{array}{ccccccc} E\ , & & Z & & & \subset & B \\ \downarrow & & \downarrow & & & & \downarrow\pi \\ \Delta_X & \subset & X\times_Y X & = & f\times f^{-1}\ \Delta_Y & \subset & X\times X \\ & & \downarrow & & & & \downarrow f\times f \\ & & \Delta_Y & & & \subset & Y\times Y \end{array} \qquad (9)$$

in which $\pi: B \to X\times X$ is the blowing-up along the diagonal Δ_X and E is the exceptional divisor and Z is the <u>residual scheme</u> of E in $\pi^{-1}\ X\times_Y X$; that is, Z is defined by the relation among ideals on B ,

$$I(Z)\ .\ I(E) = I(\pi^{-1}\ X\times_Y X)\ . \qquad (10)$$

This definition of Z is Laksov's; however, earlier Ronga had made a somewhat similar construction. Laksov then derived a residual-intersection formula, one giving the class of Z , from Grothendieck's "key" formula. Pushing it down to X , he obtained the double-point formula.

Laksov ran into technical difficulties and had to assume that no component of Z lies entirely in E . Fulton [1978] overcame these difficulties and, together, Fulton and Laksov [1978] cleaned up the treatment further using advanced in intersection theory made by Fulton and MacPherson (1978). Shortly afterwards, intersection theory entered a period of total reconstruction; with Fulton and MacPherson as principal architects. A powerful theory with a fresh point of view and great generality has now been constructed; it is presented in outline in Fulton-MacPherson [1980]. (A brief résumé, addressed to multiple-point theory, is included in Kleiman [1981b]. This theory of rational-equivalence groups and operators on them will be used freely below. Also without further comment, we shall work in the setting that seems most appropriate for multiple-point theory, the category of divisorial, universally catenary noetherian schemes and quasi-projective maps.

The version of the residual-intersection theorem that now seems best is this. Consider a diagram,

$$\begin{array}{ccccc} & & R = \mathbb{P}(I(W)) & & \\ & & \downarrow p & & \\ X & \xleftarrow{p_1} & P & \xleftarrow{q} & X \\ f\downarrow & & \downarrow p_2 & & \\ Y & \xleftarrow{h} & H & & \end{array} \qquad (11)$$

in which the square is cartesian and q is a closed embedding with ideal $I(W)$. Assume that f and p_2p and p_2q are local complete intersections, with f and p_2q of the same codimension (which need not be constant). Denote the excess in the codimension of f over that of p_2q by m, and denote the jth Chern operator of the corresponding difference of virtual normal bundles by $c_j(f/p_2q)$,

$$c_j\ (f/p_2q) = c_j\ (q^*\ p_1{}^*\ \nu_f - \nu_{p_2q})\ . \tag{12}$$

(Thus $c_m\ (f/p_2q)$ might be termed the Euler operator of the pair of maps.) Then the formulas,

$$f^*\ |H = p_*\ (p_2p)^* + q_*c_m(f/p_2q)(p_2q)^*\ , \tag{13}$$

$$p_*\ c_1(0_R(1))^k\ (p_2p)^* = -q_*\ c_{m+k}(f/p_2q)(p_2q)^* \text{ for } k \geqslant 1\ , \tag{14}$$

hold in the group of contravariant operators $A.p_2$.

This residual-intersection theorem, like earlier versions, is proved by a more or less straightforward adaptation of Laksov's original argument. The theorem is reduced by factoring f to the case that f is an embedding and then, by making blowups to the case that W and R are divisors; finally the formulas are checked against the (modern) definitions. Grothendieck's "key" formula is no longer used. In fact, a rather strong form of that formula and a rather strong form of the self-intersection formula are special cases of the excess-intersection formula, which in turn is the case of (13) in which q is an isomorphism and R is empty.

The new residual-intersection theorem brings greater simplicity and clarity to multiple-point theory. (Indeed, it was while working on Kleiman [1981b] that the author was led to it in March 1979.) The theorem is applied with $f:X \to Y$ for $h:H \to Y$ and with the diagonal for W . It is not hard to see that, in the setup of (9), the scheme Z is equal to the scheme R of (11). However, there is no longer any need for X and Y to be smooth and projective; rather, $f:X \to Y$ need only be a projective local complete intersection. (This generalization to singular varities can give results different from Johnson's generalization [1978] already when X is a curve in 3-space and f is a general projection into the plane.) Moreover, there is no longer any need to work over a field or, for that matter, over any base.

For convenience in the application, let i denote the involution of R covering the "switch" involution of $P = X\times_Y X$, and set

$$f_1 = p_2p,\quad c_k = (\nu_f) \quad \text{and} \quad t = c_1\ (0_R(1))\ . \tag{15}$$

Applying f_{1*} on the right to (13) and (14) and identifying the terms (using the general compatibility of pushout and pullback and the hypothesis that p_2q is the identity), we get

$$f_{1*}\, i_*\, f_1{}^* = f^*\, f_* - c_n \tag{16}$$

$$f_{1*}\, i_*\, t^k\, f_1{}^* = -c_{k+n} \qquad \text{for } k \geqslant 1 \,, \tag{17}$$

where n is the codimension of f. Applying (16) to the fundamental class m_1 of X, we obtain the double-point formula,

$$m_2 = f^*\, f_*\, m_1 - c_n\, m_1 \,, \tag{18}$$

where m_2 denotes the pushout under f_1 of the fundamental class of R.

While formulas for the number of triple points of a surface in 3-space were obtained in the 19th century and Roth (1930) obtained formulas for the characters of the triple curve and the number of quadruple points of plane-forms in 4-space, the first general higher-order multiple-point formula was asserted by Lashoff-Smale [1959] for a generic immersion of differentiable manifolds. Their work failed; Smale was kind enough to send (2/19/80) the author a page from his files, on which it is written, "Haefliger has pointed out to us that the formula . . . is wrong, the mistake lying on page 155, line 8 that $\sum_n = \Omega_{1\ldots1}$."

The goals of Lashoff-Smale were achieved by Herbert in [1975]. (Earlier, some work on numbers of triple points had been done following different approaches by Banchoff (1974) and by White (1975).) Independent of Herbert, Ronga worked on the problem. However, his preprint (which finally reached the author Dec. 1979) contains an error. Although the error is not hard to fix, Ronga has chosen not to publish. (The ideas though are pretty much found in Piene-Ronga [1979].) Ronga nearly obtained a version of the following weaker form of Herbert's formula, which Herbert also discusses:

$$m_{r+1} = f^*\, f_*\, m_r - r\, c_n\, m_r \,, \tag{19}$$

where m_s denotes the pushout under f_1 of the class $m_{s-1}{}'$, which is the counterpart for f_1 of m_{s-1}.

To obtain Formula (19), the <u>method of iteration</u> was used by Ronga (but not by Herbert). The method occurred independently to Salomonsen, who told the author about it (7/30/76). In the method, an $(r+1)$-fold-point formula for f is obtained from an r-fold-point formula for f_1. The method is based on the observation that a point z of R is an r-fold-point of f_1 if and only if $f_1(z)$ is an $(r+1)$-fold point of f. This observation is evident when f is an immersion, for then R is equal to complement of the diagonal in $X x_Y X$. The method also works remarkably well in the presence of ramification.

To proceed by iteration, it is necessary to know the virtual normal bundle of f_1 . If f is an immersion, then we have

$$\nu_{f_1} = i^* f_1{}^* \nu_f , \tag{20}$$

a formula which, as Salomonsen remarked, is intuitively evident on geometric grounds. Moreover, it is evident that f_1 is an immersion when f is.

The Herbert-Ronga Formula (19) is now easily obtained by induction on r . Beginning with the formula,

$$m_r' = f_1{}^* f_{1*} m_{r-1}' - (r-k) c_n (\nu_{f_1}) m_{r-1}' , \tag{21}$$

we apply $f_{1*} i_*$ and develop the terms using (16) and (20). To deal with the last term on the right, we need the formula,

$$i_* m_s' = m_s' ; \tag{22}$$

it obviously holds when f is an immersion, but it can also be proved when f is ramified.

When ramification may be present, the procedure is basically the same. However, the computations are more involved, because it is necessary to replace (20) by

$$f_1 = [i^* f_1{}^* \nu_f] (1) + [0_R - 0_R(-1)] . \tag{23}$$

(The following analogous formula can be obtained in the context of the residual-intersection theorem, provided that the virtual normal bundle of $p_2 q$ is equal to the pullback of a virtual bundle ν on H :

(24) $\nu_{p_2 p} = [(p_1 p)^* \nu_f](1) + [0_R - 0_R(-1)] + [(p_2 p)]^* \nu - ((p_2 p)^* \nu)(1)]$.)

Formula (23) was obtained by the author in March 1977 on developing ideas in Kleiman [1977] Ch. V, Sect. D. Independently, it was obtained by Roberts [1977], Sect. 5; he observed, moreover, in (6.6) that it yields the triple-point formula, but he was primarily concerned with using it to obtain some rather interesting stationary-point formulas. (Roberts in (6.3.1) obtains an appropriate case of (16) by developing ideas in Kleiman [1978] Chap. V, Sect. D; independently, this too had been done by the author in March 1977.) Formula (23) is implicit in Ronga [1973] proof of 2.7, according to Peine-Ronga [1979], comment before 3.4. In each of these derivations of (23) (and of (24)), the formula is obtained in essentially the same way, by determining explicitly the normal bundle of the appropriate blowing-up.

Multiple-point formulas of all orders can now be obtained formally. First of all, (23) yields

(25) $$f_{1*}i_*c_k(\nu_{f_1}) = c_k f_{1*}i_* + \sum_{j=0}^{k-1} [\sum_{\ell=0}^{k-j} \binom{n-j}{\ell}] c_j f_{1*}i_* t^{k-j} .$$ (25)

Secondly, because the pullback of the fundamental class of a target is always equal to the fundamental class of the source, we have

(26) $$f_1{}^*m_1 = m_1' .$$ (26)

Now all the multiple-point formulas one desires can be obtained by combining (26), (25), (22), (17) and (16) in a purely mechanical fashion.

Here are some examples:

$$m_3 = f^*f_* m_2 - 2c_n m_2 + 2(\sum_{j=0}^{n-1} 2^{n-1-j} c_j c_{2n-j}) m_1 .$$ (27)

$$n=1,\ m_4 = f^*f_* m_3 - 3c_1m_3 + 6c_2m_2 - 6(c_1c_2 + 2c_3) m_1 .$$ (28)

$$n = 2,\ m_4 = f^*f_* m_3 - 3c_2m_3 + 6(c_1c_3 + 2c_4) m_2$$ (29)
$$- 6(c_1c_2c_3 + 2c_1^2c_4 + 5c_1c_5 + 3c_2c_4 + 12c_6 + c_3^2) m_1 .$$

$$n = 3,\ m_4 = f^*f_* m_3 - 3c_3m_3 + 6(c_2c_4 + 2c_1c_5 + 4c_6) m_2$$ (30)
$$- 6(c_2c_3c_4 + 2c_2^2c_5 + 10c_1c_2c_6 + 26c_2c_7 + 3c_1c_3c_5$$
$$+ 12c_1^2c_7 + 60c_1c_8 + 9c_3c_6 + 72c_9 + 9c_4c_5) m_1 .$$

$$n = 1,\ m_5 = f^*f_* m_4 - 4c_1m_4 + 12c_2m_3 - 24(c_1c_2 + 2c_3) m_2$$ (31)
$$+ 24(c_1^2c_2 + 5c_1c_3 + 6c_4 + c_2^2) m_1 .$$

$$n = 1,\ m_6 = f^*f_* m_5 - 5c_1m_5 + 20c_2m_4 - 60(c_1c_2 + 2c_3) m_3$$ (32)
$$+ 120(c_1^2c_2 + 5c_1c_3 + 6c_4 + c_2^2) m_2$$
$$-120(c_1^3c_2 + 9c_1^2c_3 + 26c_1c_4 + 3c_1c_2^2 + 8c_2c_3 + 24c_5)m_1 .$$

Unfortunately, no general closed form is evident.

Some features do stand out. Each formula begins with the Herbert-Ronga Fomula (19); the remaining terms correct for the presence of ramification and vanish when there is none. In the expression for m_r, the coefficient of m_s contains the factor,

(33) $$(-1)^{r-s} (r-1)(r-2)\dots s ,$$ (33)

except for the first term $f^*f_*m_{r-1}$. During the course of the computation, it is not at all evident that the coefficient will contain this factor (except for the sign), and so its would-be presence serves as a useful check.

The r-fold-point formula obtained by the method of iteration is clearly valid when f is projective and r-<u>generic</u> and, for the

Herbert-Ronga Formula (19), unramified. The term "r-generic" means that each of the following r maps is a local complete intersection of the same codimension: f itself, the derived map f_{r-1}, the map f_2 derived from f_1 the way f_1 was from $f, \ldots,$ the map f_{r-1} derived from f_{r-2} the way f_1 was from f.

For example, $f: X \to Y$ is r-generic if X and Y are smooth S-schemes and f is an r-fold self-transverse immersion; "r-fold self-transverse" means intuitively that any r-branches are transverse, precisely that the tangent spaces of a (geometric) fiber of X/S at any r points in a fiber $f^{-1}y$ are in general position when embedded in the tangent space of Y/S at y. (This condition, in an appropriate form, was used by Lashoff-Smale, Herbert, and Ronga.) However, f may be an r-generic immersion without being r-fold self-transverse; consider, for instance, the difference between a tacnode and a simple node.

Suppose that f is a local complete intersection and that X or Y (and so X) is Cohen-Macaulay. Then a necessary and sufficient condition for f to be r-generic is that f be dimensionally r-generic, that is, that the derived maps $f_1, \ldots, f_{r-1}$ simply each have the same codimension as f. Moreover, R is obviously Cohen-Macaulay if f is 2-generic.

If X and Y are smooth over some base S, then of course f is a local complete intersection, but there is no reason why in general R should be smooth or f_1 should be a local complete intersection if also f_1 has the same codimension as f. However (see Roberts [1979] Thms. (0.1), (0.2) and Kleiman [1977] Ch. V, Sect. D) if f is a general central projection of a smooth variety X over an algebraically closed field of characteristic 0, then R is smooth of the appropriate dimension; a similar result holds in characteristic p but first X may have to be reembedded. Moreover (Laksov [1978] Prop. 17) in this setup, m_2 has no multiple components if $\mathrm{cod}(f) \geqslant 1$.

For the r-fold point formula to be valid, the condition that f be r-generic is really too strong. Call f practically r-generic if f is a local complete intersection and if there exists a closed subset S of Y such that the restriction of f over the complement of S is r-generic and such that the restriction of f over the complement of S is r-generic and such that the codimension of $f^{-1}S$ in X is greater than $(r-1)n$ where (as always) n is the codimension of f. The formula is valid when f is practically r-generic because then each component of m_r has codimension $(r-1)n$ exactly. (If n is only locally constant, replace $(r-1)n$ by the function whose value at x in X is the maximum value of the sum $n(x_2) + \ldots + n(x_r)$ as $x_2,\ldots,x_r$ range over the fiber $f^{-1}f(x)$.)

Suppose f has an $\overline{S}_2$-singularity x; that is, x is a (geometric) point of X such that $\Omega_f^1(x)$ has dimension at least 2. Then the fiber $f_1^{-1}(x)$ has dimension at least 1, and so the fiber $(f_1\ldots f_{r-1})^{-1}(x)$ has dimension at least $r-1$. Suppose also that f is a local complete intersection of codimension n. Then the locus of $\overline{S}_2$-singularities has codimension at most $2(n+2)$ in X. It follows that, if f is practically r-generic, then r and n are limited to the following range: $r=2$ or 3 and n arbitrary, $r=4$ and $n=1,2,3$, and $r=5$ or 6 and $n=1$. The multiple-point formulas in this range are just the ones given above, (18), (27)-(32). Outside this range, of course, the multiple-point formulas obtained by iteration are valid for a practically r-generic map with no $\overline{S}_2$=singularity, just as the Herbert-Ronga Formula (19) is valid for a practically r-generic immersion, a map with no $\overline{S}_1$-singularity.

The presence of the factor (33) is a defect of the method of iteration. This defect arises because, for each r-fold point x of f, the $r-1$ points $x_2,\ldots,x_r$ with the same image as x are taken implicitly in the enumeration with a specific order in each of the $(r-1)!$ ways this is possible. The defect is especially serious in differential topology; Herbert [1975] Rem. p. 9 points out that m_r must vanish for $r \geqslant 3$ unless X and Y are oriented and n is even because $\mathbb{Z}_2$-coefficients are necessary unless this is so. Herbert overcame the difficulty (which is present in the Lashoff-Smale theory) and obtained a refined version of the Herbert-Ronga Formula (19). Ronga [1980] recently gave a different and shorter proof of the refined formula; this proof uses the symmetric product and a generalization of a result of Quillen, which is a version of the excess-intersection formula -- the case of residual-intersection formula (13) in which q is an isomorphism (note that n is not necessarily constant).

The residual-intersection formula (13) led the author, who was aware of Herbert's work but not Ronga's, in May 1979 to a version of Herbert's formula and a corresponding refined version of the triple-point formula (27). These two formulas will now be discussed; a detailed treatment will be found in Kleiman [1982].

The refined formulas involve three classes t_r, u_r, v_r, each refining m_r in a different way. Their definitions involve the Hilbert scheme $H = H_r$, which parametrizes the length-r subschemes in the fibers of the projective map f. Let $h:H \to Y$ denote the structure map, $W = W_r$ the universal family, which is a closed subscheme of the product $P = X \times_Y H$, and $R = R_r$ the residual scheme of W in P. Thus, we have the setup of diagram (11); use p_1, p_2, p and q correspondingly.

Define the refined classes as the following pushouts of fundamental classes:

$$t_r = h_*[H], \quad u_r = (p_1q)_*[W], \quad v_{r+1} = (p_1p)_*[R] . \tag{34}$$

Thus t_r is a class on Y representing the points whose fibers contain length r-subschemes; it is analogous to Herbert's transverse class. The class u_r on X represents the points which are parts of length-r subschemes of the fibers of f; it is a version of m_r without the factor $(r-1)!$.. Finally, v_{r+1} is a class on X representing the points x such that there exists a length-(r+1) subscheme of the fiber $f^{-1}f(x)$ that is an extension of some length-r subscheme by x; in fact, R parametrizes a universal (flat) family of length-1 extensions in the fibers of f of the family W/H.

The various classes are related as follows. First of all, we have

$$f_* u_r = rt_r \tag{35}$$

because $p_2q:W \to H$ is finite, flat and of degree r. Now, in a number of cases, we have

$$v_{r+1} = u_{r+1} \tag{36}$$

$$m_r = (r-1)!\, u_r . \tag{37}$$

Indeed, if $r=1$, then both sides of (37) reduce to the fundamental class of X. If $r = 1$, then (36) holds, because R_1 and W_2 are canonically isomorphic; both schemes parametrize the length-1 extensions of the length-1 subschemes of the fibers of f. If $r = 2$, then (37) holds because $H_1 = X$ and so $m_2 = v_2$ and because $v_2 = u_2$ as was just observed. If f is practically 3-generic, then it can be proved that (36) holds with $r=2$ and (37) holds with $r=3$.

If f is unramified, then (36) and (37) hold for every r. Indeed, then P/H is unramified, so W/H is étale. Hence W is open and closed in P. Therefore R is equal to the complement of W in P. Consequently, the natural map from R_r onto W_{r+1} is an isomorphism. Thus (36) holds. Now, it is not hard to prove by induction on r that, since f is unramified, X_r, the source of f_r, is equal to the complement of all diagonals in the r-fold direct self-product of X/Y and that there is a natural étale map of degree $(r-1)!$ from X_r to W_r. Thus (37) holds.

The refined formulas are these:

$$u_{r+1} = f^*t_r - c_n u_r, \quad f \text{ unramified}; \tag{38}$$

$$v_3 = f^*t_2 - c_n u_2 + (\textstyle\sum_{j=0}^{n-1} s^{n-1-j} c_j c_{2n-j}) u_1 . \tag{39}$$

The formulas hold when f and p_2p are local complete intersections of codimension n. This condition is weaker than r-genericity; there is no restriction on the size of the locus of s-fold points for $s < r$, and the restriction on the size of the locus of r-fold points is relative to the size of the locus of (r-1)-fold points (more precisely, p_2p should have codimension n). If H is Cohen-Macaulay and f is a local complete intersection, then p_2p will be a local complete intersection if it has the same codimension n as f. Of course, the condition that p_2p be a local complete intersection of codimension n can be relaxed a little to the condition that there exist a closed subset S of Y such that $h^{-1}S$ is nowhere dense in H and the restriction of p_2p over $h^{-1}S$ is a local complete intersection of codimension n. Note also that (38) and (39) are stronger than (19) and (27); term by term, (19) is $r!$ times (38) and the right side of (27) is twice that of (39), while $m_3 = 2v_3$ when f is practically r-generic.

Formula (38) follows from the residual-intersection formula (13), by applying p_{1*} on the left and $[H]$ on the right and identifying the terms, noting that $c_m(f/p_2q)$ is equal to $c_n(\nu_f)$ because p_2q is étale when f is unramified.

Formula (39) is obtained similarly. However, the term involving $c_m(f/p_2q)$ requires more work to identify. Set

$$w = p_2q, \quad E = w_*O_W, \quad L = \Lambda^2 E \tag{40}$$

Since w is finite and E is locally free of rank 2, it follows that there is a canonical embedding of W in $\mathbb{P}(E)$ as a divisor with ideal isomorphic to $(w^*L^2)(-2)$. Hence, w is a local complete intersection, its virtual normal bundle is equal to the pullback of $(L^{-2} - L^{-1})$, and we have

$$c(\nu_w)^{-1} = w^*(1-c_1(L)) \,/\, (1-2c_1(L)) . \tag{41}$$

Now, the isomorphism between R_1 and W_2 identified $O(1)$ and w^*L. By transferring the term in question to R_1 and using (17), the term is easily identified, completing the derivation of (39).

The derivation of (39) involves less mechanical computation than that of (27), and the shape of the final expression is evident sooner; for instance, the presence of the coefficient 2^{n-1-j} in (39) is a priori evident from (41), whereas its presence in (27) seems to be an unforeseeable combinatorial quirk. Thus hope is raised that the method of derivation of (39) may eventually lead not only to an assertion of the existence of refined versions of all the formulas obtainable by iteration but also to explicit closed forms. The first obstacle to further progress is the fact that W_3/H_3 is not in general a local complete intersection; future work may yield a version of the residual-intersection theorem appropriate to this case, perhaps one involving Segre classes of W/H, or future work may reveal a better choice for H than the Hilbert scheme.

Future work in multiple-point theory should also be devoted to formulas enumerating the sequences of r points with the same image and, perhaps, with one or more singled out, such that n_1 of the r points lie infinitely close, n_2 others of them lie infinitely close, etc. with $n_1 + n_2 + \ldots = r$. This work would enlarge on Robert's work [1979] and should incorporate a theory of multiple-coincidence formulas à la Schubert [1879], Ch. V, a theory of Thom polynomials, and applications to enumerative geometry, special divisors, etc. A survey of some work in this direction is found in Kleiman [1976], pp. 467-68, and through Kleiman [1977]. More recent work includes Arbarello et al. [1980], Canuto [1979], Grayson [1979], Griffiths-Harris [1979], Holme [1978], [1979], Kleiman [1981a], Laudal [1978], Le Barz [1978], [1979a], [1979b], [1980], [1981b], [1981c], Piene [1977], [1978a], [1978b], [1979], [1980], Piene-Ronga [1979], Roberts-Speiser [1980], Vainsencher [1980], von zur Gathen [1980], Urabe [1980]. Thus, multiple-point theory is a lovely old and wide field, in which there is much activity now but in which there is much yet to do.

Footnotes

[1] This manuscript was originally prepared in connection with the celebration of the 10th anniversary of the founding of the Universidad Simon Bolívar, Caracas, which was held January 28-31, 1980. The original manuscript appeared in Vol. 3, Geometria algebrica y teoria de numeros, A. B. Altman coordinator, of the Simposia series of the university. The present manuscript is a reproduction of the original one with minor changes only.

[2] J. S. Guggenheim Fellow.

References

Arbarello, E., Cornalba, M., Griffiths, P., Harris. : Geometry of Algebraic Curves, to appear.

Canuto, G. [1979]: Associated Curves and Plücker Formulas in Grassmannians. Inventiones math. 53, 77-90 (1979).

Catanese, F. [1979]: On Severi's Proof of the Double Point Formula. Comm. in Alg. 7 (7), 763-773 (1979).

Fulton, W. [1978]: A NOTE ON RESIDUAL INTERSECTIONS AND THE DOUBLE POINT FORMULA. Acta math. 140, 93-101 (1978).

Fulton, W., MacPherson [1980]: BIVARIANT THEORIES. Preprint, Brown Univ. Providence (1980).

Fulton, W., Laksov, D. [1977]: Residual intersections and the double-point formula. Real and complex singularities, Oslo 1976, P. Holm, ed., pp. 171-177. Sijthoff & Noordhooff (1977).

Grayson, D. [1979]: COINCIDENCE FORMULAS IN ENUMERATIVE GEOMETRY. Comm. in Alg. 7(16), 1685-1711 (1979).

Griffiths, P., Harris, J. [1979]: Algebraic Geometry and Local Differential Geometry. Ann. Ec. N.S. 12(3), 355-452 (1979).

Herbert, R. [1975]: MULTIPLE POINTS OF IMMERSED MANIFOLDS. Thesis, Univ. of Minnesota (1975), Amer. Math. Soc. Memoire No. 250 (Oct. 1981).

Holme, A. [1978]: Deformation and Stratification of Secant Structure. Algebraic Geometry, Proceedings, Tromø, Norway 1977, L. Olson, ed., pp. 60-91. Lecture Notes inMath., 687. Springer (1978).

_____[1979]: On the dual of a smooth variety. Algebraic Geometry, Proceedings, Copenhagen 1978, K. Lonsted, Ed., pp. 144-156. Lecture Notes in Math., 732. Springer [1979).

Johnson, K. [1978]: Immersion and Embedding of Projective Varieties. Acta math., 140, 49-74 (1978).

Kleiman, S. [1976]: Problem 15. Rigorous foundation of Schubert's enumerative calculus. Mathematical Developments arising from Hilbert Problems, Proc. Symposia Pure Math., Vol. XXVIII, pp. 445-82. A.M.S. (1976).

_____[1977]: The enumerative theory of singularities. Real and complex singularities, Oslo 1976, P. Holm, ed., pp. 297-396. Sijthoff & Noordhooff (1977).

_____[1981a]: Concerning the dual variety. 18th Scand. Congress Math. Proceedings, Birkhäuser (1981), 386-396.

_____[1981b]: Multiple-point formulas I: Iteration. Acta Math. 147, 13-49 (1981).

_____[1982]: Multiple-point formulas II: O-cycles. In preparation.

_____[1980]: RELATIVE DUALTIY FOR QUASI-COHERENT SHEAVES. Compositio Math. 41 (1), 39-60 (1980).

Laksov, D. [1978a]: RESIDUAL INTERSECTIONS AND TODD'S FORMULA FOR THE DOUBLE LOCUS OF A MORPHISM. Acta Math., 140, 75-92 (1978).

_____[1978b]: Secant bundles and Todd's formula for the double points of maps into $\mathbb{P}^n$. Proc. Lond. Math. Soc. 37, 120-142 (1978).

Lashof, R., Smale, S. [1958]: Self-intersections of immersed manifolds. J. Math. Mech. 8, 143-157 (1959).

Laudal, O. [1978]: A Generalized Trisecant Lemma. Algebraic Geometry, Proceedings, Tromsø, Norway 1977, L. Olson, ed., pp. 112-149. Lecture Notes in Math., 687. Springer (1978).

LeBarz, P. [1978]: Géometrie énumérative pour les multisécantes. Variétés analytic compactes, Nice 1977, pp. 116-167. Lecture Notes in Math., 687. Springer (1978).

_____[1979a]: Validité de certaines formules de géometrie énumérative. C.R.A.S. Paris. 289, 755-758 (1979).

_____[1979b]: Courbes générales de $\mathbb{P}^3$. Math. Scand. 44, 243-277 (1979).

_____[1980]: Une courbe gauche avec -4 quadrisécantes. Preprint Univ. Nice (1980).

_____[1981a]: Formules pour les multisécantes des surfaces, C.R.A.S. 292, 797-800 (1981).

_____[1881b]: Quelques calculs dans $\mathrm{Hilb}^k\,\mathbb{P}^N$ et la variété des alignements, in preparation.

_____[1981c]: Une Application de la Formule de Fulton-MacPherson. Preprint Univ. Nice (June 1981).

Piene, R. [1977]: Numerical characters of a curve in projective n-space. Real and complex singularities, Oslo 1976, P. Holm, ed., pp. 475-496. Sijthoff & Noordhooff (1977).

_____[1978a]: Polar classes of singular varieties. Ann. S. Ec. N. Sup. 11, 274-276 (1978).

_____[1978b]: Some Formulas for a Surface in $\mathbb{P}^3$. Algebraic Geometry, Proceedings, Tromsø, Norway 1977, L. Olson, ed., pp. 196-235. Lecture Notes in Math, 687. Springer (1978).

_____[1979]: A proof of Noether's formula for the arithmetic genus of an algebraic surface. Compositio math. (1) 38, 113-119 (1979).

_____[1980]: CUSPIDAL PROJECTIONS OF SPACE CURVES. Preprint, Univ. of Oslo, Norway, Feb. (1980), Math. Ann. 256, 95-119 (1981).

Piene, R., Ronga, F. [1979]: A GEOMETRIC APPROACH TO THE ARITHMETIC GENUS OF A PROJECTIVE MANIFOLD OF DIMENSION THREE. Preprint July (1979), Topology, 20, 179-190 (1981).

Roberts, J. [1979]: Some properties of double point schemes. Preprint May (1979), Compositio Math., 41(1), 61-94 (1980).

Roberts, J., Speiser, R., [1980]: Schubert's enumerative geometry of triangles from a modern viewpoint. Algebraic Geometry, Proceedings, Chicago Circle (1980), A. Libgober and P. Wagreich, ed., pp. 272-281, Lecture Notes in Math, 862. Springer (1981).

Roberts, J., Speiser, R., [1980]: Enumerative geometry of triangles. In preparation.

Ronga, F. [1973]: Le calcul des classes duals aux points doubles d'une application. Compositio math. (2)27, 223-232 (1973).

_____[1980]: On multiple points of smooth immersions. Preprint March (1980).

Schubert, H. [1879]: Kalkül der abzählenden Geometrie. Reprint. Springer (1979).

Vainsencher, I. [1980]: COUNTING DIVISORS WITH PRESCRIBED SINGULARITIES. To appear in Proc. A.M.S.

von zur Gathen, J. [1980]: SEKANTENRÄUME VON KURVEN. Inaugural-Dissertation, Univ. Zurich (1980).

Urabe, T. [1980]: Duality of numerical characters of polar loci. Preprint March (?) (1980).

PROGRESS IN MATHEMATICS

Already published

PM 1 Quadratic Forms in Infinite-Dimensional Vector Spaces
Herbert Gross
ISBN 3-7643-1111-8, 432 pages, paperback

PM 2 Singularités des systèmes différentiels de Gauss-Manin
Frédéric Pham
ISBN 3-7643-3002-3, 346 pages, paperback

PM 3 Vector Bundles on Complex Projective Spaces
C. Okonek, M. Schneider, H. Spindler
ISBN 3-7643-3000-7, 396 pages, paperback

PM 4 Complex Approximation, Proceedings, Quebec, Canada, July 3-8, 1978
Edited by Bernard Aupetit
ISBN 3-7643-3004-X, 128 pages, paperback

PM 5 The Radon Transform
Sigurdur Helgason
ISBN 3-7643-3006-6, 202 pages, paperback

PM 6 The Weil Representation, Maslov Index and Theta Series
Gérard Lion, Michèle Vergne
ISBN 3-7643-3007-4, 348 pages, paperback

PM 7 Vector Bundles and Differential Equations
Proceedings, Nice, France, June 12-17, 1979
Edited by André Hirschowitz
ISBN 3-7643-3022-8, 256 pages, paperback

PM 8 Dynamical Systems, C.I.M.E. Lectures, Bressanone, Italy, June 1978
John Guckenheimer, Jürgen Moser, Sheldon E. Newhouse
ISBN 3-7643-3024-4, 298 pages, paperback

PM 9 Linear Algebraic Groups
T. A. Springer
ISBN 3-7643-3029-5, 314 pages, hardcover

PM10 Ergodic Theory and Dynamical Systems I
A. Katok
ISBN 3-7643-3036-8, 346 pages, hardcover

PM11 18th Scandinavian Congress of Mathematicians, Aarhus, Denmark, 1980
Edited by Erik Balslev
ISBN 3-7643-3040-6, 526 pages, hardcover

PM12 Séminaire de Théorie des Nombres, Paris 1979-80
Edited by Marie-José Bertin
ISBN 3-7643-3035-X, 404 pages, hardcover

PM13 Topics in Harmonic Analysis on Homogeneous Spaces
Sigurdur Helgason
ISBN 3-7643-3051-1, 152 pages, hardcover

PM14 Manifolds and Lie Groups, Papers in Honor of Yozô Matsushima
Edited by J. Hano, A. Marimoto, S. Murakami, K. Okamoto, and H. Ozeki
ISBN 3-7643-3053-8, 476 pages, hardcover

PM15 Representations of Real Reductive Lie Groups
David A. Vogan, Jr.
ISBN 3-7643-3037-6, 776 pages, hardcover

PM16 Rational Homotopy Theory and Differential Forms
Phillip A. Griffiths, John W. Morgan
ISBN 3-7643-3041-4, 258 pages, hardcover

PM17 Triangular Products of Group Representations and their Applications
S.M. Vovsi
ISBN 3-7643-3062-7, 142 pages, hardcover

PM18 Géométrie Analytique Rigide et Applications
Jean Fresnel, Marius van der Put
ISBN 3-7643-3069-4, 232 pages, hardcover

PM19 Periods of Hilbert Modular Surfaces
Takayuki Oda
ISBN 3-7643-3084-8, 144 pages, hardcover

PM20 Arithmetic on Modular Curves
Glenn Stevens
ISBN 3-7643-3088-0, 236 pages, hardcover

PM21 Ergodic Theory and Dynamical Systems II
A. Katok, editor
ISBN 3-7643-3096-1, 226 pages, hardcover

PM22 Séminaire de Théorie des Nombres, Paris 1980-81
Marie-José Bertin, editor
ISBN 3-7643-3066-X, 374 pages, hardcover

PM23 Adeles and Algebraic Groups
A. Weil
ISBN 3-7643-3092-9, 138 pages, hardcover

PROGRESS IN PHYSICS

Already published

PPh1 Iterated Maps on the Interval as Dynamical Systems
Pierre Collet and Jean-Pierre Eckmann
ISBN 3-7643-3026-0, 256 pages, hardcover

PPh2 Vortices and Monopoles, Structure of Static Gauge Theories
Arthur Jaffe and Clifford Taubes
ISBN 3-7643-3025-2, 294 pages, hardcover

PPh3 Mathematics and Physics
Yu. I. Manin
ISBN 3-7643-3027-9, 112 pages, hardcover

PPh4 Lectures on Lepton Nucleon Scattering and Quantum Chromodynamics
W.B. Atwood, J.D. Bjorken, S.J. Brodsky, and R. Stroynowski
ISBN 3-7643-3079-1, 574 pages, hardcover

PPh5 Gauge Theories: Fundamental Interactions and Rigorous Results
P. Dita, V. Georgescu, R. Purice, editors
ISBN 3-7643-3095-3, 406 pages, hardcover